Oleksandr Motsak

Non-commutative Computer Algebra with applications

Oleksandr Motsak

Non-commutative Computer Algebra with applications

Graded commutative algebra and related structures in Singular with applications

Südwestdeutscher Verlag für Hochschulschriften

Impressum/Imprint (nur für Deutschland/only for Germany)
Bibliografische Information der Deutschen Nationalbibliothek: Die Deutsche Nationalbibliothek verzeichnet diese Publikation in der Deutschen Nationalbibliografie; detaillierte bibliografische Daten sind im Internet über http://dnb.d-nb.de abrufbar.
Alle in diesem Buch genannten Marken und Produktnamen unterliegen warenzeichen-, marken- oder patentrechtlichem Schutz bzw. sind Warenzeichen oder eingetragene Warenzeichen der jeweiligen Inhaber. Die Wiedergabe von Marken, Produktnamen, Gebrauchsnamen, Handelsnamen, Warenbezeichnungen u.s.w. in diesem Werk berechtigt auch ohne besondere Kennzeichnung nicht zu der Annahme, dass solche Namen im Sinne der Warenzeichen- und Markenschutzgesetzgebung als frei zu betrachten wären und daher von jedermann benutzt werden dürften.

Verlag: Südwestdeutscher Verlag für Hochschulschriften GmbH & Co. KG
Dudweiler Landstr. 99, 66123 Saarbrücken, Deutschland
Telefon +49 681 37 20 271-1, Telefax +49 681 37 20 271-0
Email: info@svh-verlag.de

Zugl.: Kaiserslautern, TU, Diss., 2010

Herstellung in Deutschland:
Schaltungsdienst Lange o.H.G., Berlin
Books on Demand GmbH, Norderstedt
Reha GmbH, Saarbrücken
Amazon Distribution GmbH, Leipzig
ISBN: 978-3-8381-2752-1

Imprint (only for USA, GB)
Bibliographic information published by the Deutsche Nationalbibliothek: The Deutsche Nationalbibliothek lists this publication in the Deutsche Nationalbibliografie; detailed bibliographic data are available in the Internet at http://dnb.d-nb.de.
Any brand names and product names mentioned in this book are subject to trademark, brand or patent protection and are trademarks or registered trademarks of their respective holders. The use of brand names, product names, common names, trade names, product descriptions etc. even without a particular marking in this works is in no way to be construed to mean that such names may be regarded as unrestricted in respect of trademark and brand protection legislation and could thus be used by anyone.

Publisher: Südwestdeutscher Verlag für Hochschulschriften GmbH & Co. KG
Dudweiler Landstr. 99, 66123 Saarbrücken, Germany
Phone +49 681 37 20 271-1, Fax +49 681 37 20 271-0
Email: info@svh-verlag.de

Printed in the U.S.A.
Printed in the U.K. by (see last page)
ISBN: 978-3-8381-2752-1

Copyright © 2011 by the author and Südwestdeutscher Verlag für Hochschulschriften GmbH & Co. KG and licensors
All rights reserved. Saarbrücken 2011

Contents

Preface   iii

1 Algebraic preliminaries and notations   1
   1.1 Algebras over fields 1
   1.2 Modules over algebras 2
   1.3 Noetherian modules 4
   1.4 Free modules over algebras 5
   1.5 Graded structures 6
   1.6 Tensor algebra 9
   1.7 Binary relations 12

2 Computer Algebra preliminaries   15
   2.1 Standard monomials and monomial orderings 15
   2.2 G-algebras 18
   2.3 Definitions of Gröbner Bases in quotient algebras 22
   2.4 Usual approach to computations in factor algebras 23

3 Gröbner bases in graded commutative algebras   25
   3.1 Green's approach 25
   3.2 Preliminaries 31
   3.3 Direct approach 32
   3.4 Characterizations of Gröbner Bases 40
   3.5 Criteria 43
   3.6 Kernel and preimage of a graded homomorphism 44

4	Localization			47
	4.1	The commutative localization		47
	4.2	Non-commutative localization		49
		4.2.1	Universal construction	50
		4.2.2	Ore construction	50
	4.3	Central localization		52
	4.4	Rings Associated to Monomial Orderings		54
5	Syzygies and free resolutions			59
	5.1	Computer Algebra for modules		59
	5.2	Assumptions on orderings		66
	5.3	Schreyer ordering and syzygies of leading terms		68
	5.4	Computation of a free resolution		75
6	Graded commutative algebras in SINGULAR			85
	6.1	High level interface - users manual		85
	6.2	Product of monomials in graded commutative algebras		95
	6.3	Detection of a graded commutative structure		96
	6.4	A bit about SINGULAR internals		96
	6.5	Implementing an induced ordering after Schreyer		97
7	Applications			103
	7.1	Projective Geometry		103
		7.1.1	Introduction to sheaf cohomology	104
		7.1.2	Exterior algebra method for sheaf cohomology computation	109
		7.1.3	Sheaf cohomology: benchmarks	116
	7.2	Coordinate-free verification of affine geometry theorems		125
	7.3	Super-symmetry		132
	7.4	Cohomology rings of finite p-groups		136
8	Conclusion and Future Work			137
	Bibliography			139

Preface

Overview

The commutative polynomial algebra over a field plays the main role in Computational Commutative Algebra. On the other hand, its Koszul dual algebra, known as exterior algebra and Grassmann algebra describes for instance, the geometry of a Euclidean vector space.

In the thesis we have developed a full Computer Algebra framework, naturally unifying these both algebras in the class of polynomial graded commutative algebras.

Recall that a *super-commutative algebra* is a $\mathbb{Z}_2$-graded algebra, where the product satisfies

$$b \cdot a = (-1)^{|a||b|} a \cdot b,$$

for all ($\mathbb{Z}_2$-graded) homogeneous elements a, b (the grading, i.e. degree, of an element a is denoted by $|a| \in \mathbb{Z}_2$).

Algebraic structures which super-commute in the above sense are sometimes referred to as *skew-commutative associative algebras* to emphasize the anti-commutation, or, to emphasize the grading, *graded commutative* or, even *commutative*, provided the super-commutativity is understood (e.g. in advanced physics).

Nowadays algebras with super-commutative structure are becoming more and more important through their use in a wide range of applications, including theoretical physics and various branches of mathematics (some examples are given below). Although one can treat them as GR-algebras (cf. [5, 86]) or solvable polynomial rings (cf. [78, 76]), we introduce the computer algebra notions directly for these algebras and give rigorous proofs with an emphasis on differences occurring due to the presence of zero divisors. In particular, we show that graded commutative algebras admit such a characterization of a Gröbner basis (cf. Theorem 3.4.6) which leads to a particularly efficient generalization of Buchberger's algorithm (cf. [16, 67]).

In our treatment we try to follow, as far as possible, the Computer Algebra approach developed in [67].

Furthermore, we show that central localizations of finitely generated non-commutative algebras can be tackled, similarly to localizations of commutative polynomial algebras (cf. [67]), by considering straightforward generalizations of rings associated to orderings.

The central notion of Computer Algebra for polynomial rings being Groebner bases, while nowadays the notions of syzygy and free resolution are becoming more and more important. In fact, there is a growing body of qualitative results and conjectures relating the geometry of a projective variety to the form of the syzygies (the minimal free resolution) of its homogeneous coordinate ring. From a practical point of view, syzygy computations play a fundamental role when it comes to experiments in algebraic geometry and to the testing of conjectures. Furthermore, syzygy computations are central to many advanced algorithms on which the experiments are based.

In principle, computing syzygies may be thought of as a byproduct of computing Gröbner bases. The *effective* computation of syzygies, however, requires the choice and fine tuning of special data structures, monomial orders, and strategies. Depending on the strategy, free resolutions can be computed just by iterating the syzygy computation (e.g. due to Schreyer), or by using one of the more sophisticated algorithms (e.g. due to La Scala and Stillman).

Even more involved is the issue of minimality (for homogeneous input) of the computed free resolution. Also here, different strategies are possible.

SINGULAR[68] provides comprehensive and considerably fast routines for computing syzygies and free resolutions over $\Bbbk[x_0, \ldots, x_n]$, where $\Bbbk$ is a field, and localizations thereof. In addition, it offers a basic syzygy (free resolution) algorithm which works for a large class of non-commutative algebras. That algorithm has been automatically improved, due to our implementation of Buchberger's algorithm over graded commutative algebras.

The current implementation of the algorithms of Schreyer and La Scala, being implemented in the SINGULAR kernel more than 10 years ago, are by far not optimized yet. Moreover, they have suffer from severe implementational issues, which, for instance, disallow us to extend them to the non-commutative case. That is why we provide an experimental prototype for a variation of these algorithms, which is designed to work seamlessly over factors of non-commutative algebras, e.g. over graded commutative algebras.

All this allows us to devise efficient algorithms for computation over graded commutative algebras and, in particular, over super-commutative algebras. Our framework has plenty of applications including, among others, automatic geometrical theorem (coordinate-free) verification, investigation of properties of exterior differential systems.

The framework developed in the scope of this thesis has proved to be efficient, by particularly efficiency demanding applications, including the classification of (graded commutative) cohomology rings of finite p-groups by D. Green and S. King (cf. Section 7.4) and the computation of sheaf cohomology due to the constructive *Bernstein-Gelfand-Gelfand correspondence* via the algorithm of Eisenbud-Fløystad-Schreyer (standalone implementation in Singular: [31] and in particular in via the GAP package HomAlg, cf. [9]).

Motivation for graded commutative and super-commutative algebras

As has been already noted above, graded commutative algebras unify the standard commutative polynomial and exterior algebras. In particular, the $\mathbb{Z}_2$-graded commutative algebras are also known as *super-commutative algebras*.

The super-commutative algebras were first introduced as symmetric algebras of super-manifolds in theoretical physics, or more precisely, super-symmetry, which is part of the theory of elementary particles and their interactions (cf. [92, 13]), where these algebras enable one to join particles with Bose-Einstein statistics and Fermi-Dirac statistics into single multiplets, and also enables one to join the internal and dynamic symmetries of gauge theories in a single super-group (for a brief account on the super-mathematics see Section 7.3).

The first appearance of super-symmetry is the Grassmann's definition of an algebra that, although non-commutative, is commutative up to a sign factor. The Grassmann algebra of a vector space is perhaps the earliest example of a grade commutative algebras. It appears extensively in Topology (e.g. cohomology algebras, Steenrod algebras, and Hopf algebras), and in Geometry (e.g., the *de Rham complex*; cf. also *de Rham cohomology*).

After being endowed with a certain differential operator graded commutative algebras become *Batalin-Vilkovisky algebras* which are used in theoretical physics for determining the ghost structure for theories, such as gravity and super-gravity, whose Hamiltonian formalism has constraints not related to a Lie algebra action. Furthermore, together with a specially defined Poisson bracket, they turn into *Gerstenhaber algebras*.

In Intersection Theory, the intersection semi-lattice of a *hyperplane arrangement* determines a combinatorial invariant of the arrangement, the *Orlik-Solomon algebra*, is a factor of a graded commutative algebra (together with the usual boundary operator δ on anti-commutative part) by certain ideal, defined due to the hyperplane arrangement (cf. [38]).

From the other side, works by G. C. Rota (cf. [69, 110]) showed the significant importance of the so called *letterplace super-algebra*, denoted by $\text{Super}(L|P)$, which can be constructed as a quotient of a super-commutative algebra (sometimes in infinite number of generators). We expect that the treatment due to [80] can be applied to these algebras as well.

Exterior differential system (EDS) (cf. [72, 71, 70]) is a graded submodule of the module of exterior differential forms on a manifold. The following properties may be shown by checking ideal membership, at each point on the manifold:

- An EDS S is *closed* if $dS \subset \langle S \rangle$, where d is the exterior derivative.
- A vector v is an *isovector* or *symmetry generator* of S if $\imath_v S \subset \langle S \rangle$, where $\imath_v$ is the interior product.

Overview of previous results and related works

Good general textbooks on commutative Computer Algebra are [1, 10]. Its use for the needs of Algebraic Geometry can be found in many textbooks, e.g. [67, 27, 126, 30]. Note that we try to follow the CA approach due to [67] and [86] as close as possible.

Ring and ideal theory of non-commutative Noetherian domains have been studied for example in [24, 2, 94, 59, 101, 83, 103, 104, 84]. Most proofs of GB computation termination rely on the basic criterion provided by [33].

Since each k-algebra is a quotient of a free associative non-commutative algebra one could do general non-commutative Computer Algebra with them in order to tackle such quotients in a general fashion (it has been studied among others in [14, 97, 99, 98, 26, 3, 63, 89]). Free non-commutative polynomial rings modulo some non-commutative GBs have been studied in [5]. Furthermore, the book [89] also gives a good theoretical account on non-commutative CA for general k-algebras with zero-divisors.

The following very popular class of non-commutative algebras (with *PBW* bases), has a lot of names. It was probably invented by J. Apel under the name "G-algebras" in [5], and studied/reinvented by many people under many different names.

They are know as algebras of solvable type in [45], A. Kandri-Rody and V. Weispfenning (cf. [76]) were probably the first to introduce non-commutative computer algebra over solvable polynomial rings and algebras directly.

These algebras are known as *PBW rings* in [19]. H. Kredel (cf. [78]) has generalized solvable polynomial rings and algebras even further by allowing scalars to be non-commutative, so that the resulting algebras may fail to be k-algebras. V. Levandovskyy together with H. Schönemann have developed a framework for computations over them in SINGULAR, where they are known as G- and GR-algebras (cf. [86, 87]).

Polynomial graded commutative algebras (in particular, exterior algebras) as well as Clifford algebras are quotients, and belong to the above class of (almost commutative) algebras.

Because of its uses in many contexts the exterior algebra is extremely popular (consider for example the list of alternative names below) as a subject for research and thus there exists a variety of possible approaches. The algebra was introduced by H. Grassmann for his approach to affine geometry (cf. [60]). His approach was followed by D. Fearnley-Sander together with T. Stockes (see Section 7.2), who were probably the first to develop computer algebra over exterior algebras (cf. [51, 123]).

Computer algebra over Clifford algebras (and in particular over exterior algebras) was done by Harley and Tuckey (cf. [72]) who considered them as particular GR-algebras. Similarly, D. Green considered graded commutative algebras as quotients of his (anti-commutative polynomial) Θ-algebras (which are simple G-algebras) by the two-sided ideal generated by squares of odd variables (cf. [62] and our overview of his results in Section 3.1).

Somewhat different accounts on the topic give: E. Green considers non-commutative k-algebras with zero-divisors with multiplicative bases (e.g. path algebras) and introduces right GBs for a class of modules (cf. [65, 64]), Th. Nuessler and H. Schoenemann tackle zero-divisors by considering additional augmentation sets (cf. [102]) and K. Madlener together with B. Reinert use saturating sets for the same purpose (cf. [91]).

We use the algorithmic treatment of commutative localizations given in [67] as a reference for our treatment of non-commutative localizations, which have been also discussed in [82, 25].

It was already noted that syzygy and free resolution computations are deeply related to GB computation (cf. [116, 117]). Our starting point for syzygy computations is the idea that syzygies of leading terms (w.r.t. some special ordering) give rise to syzygies of whole elements, which leads to a syzygy-driven Buchberger's algorithm (cf. [4, 113, 114, 96, 120, 107] and somewhat [20]).

For our Schreyer free resolution we used the results by R. La Scala and M. Stillman (cf. [79, 81]).

The most theoretically involved applications of our methods are Projective Geometry-related. Please refer to the classical texts [73, 115] and [57] for basic definitions. An excellent introduction to computational methods was given by M. Stillman at the Arizona Winter School in Tucson, March 2006, (notes [122] and videos are available online at http://math.arizona.edu/~swc/aws/06/06Notes.html).

Traditional methods for computing with sheaves and sheaf cohomology can be found in [115], [126] and [119].

The exterior algebra method for computing sheaf cohomology relies on a constructive version of BGG correspondence given in [36] and [29]. Note that the BGG correspondence (cf. [15, 11]) is a particular case of Koszul duality (cf. [53, 52, 12]).

The implementation of this method by W. Decker, D. Eisenbud and F.-O. Schreyer as package BGG (cf. [29]) in M2 (cf. [61, 126]) was the fastest known up until now.

Generalization of this method allows one to compute higher direct images of sheaves (cf. [39]). Apart from constructing Beilinson monad and Horrocks-Mumford bundle, this approach has been used for plenty interesting theoretical applications, e.g. computation of resultants and Chow forms (cf. [35, 42]), cohomologies of hyperplane arrangement (cf. [38]), and others (e.g. [43]).

This method and most of its applications, for example the investigation of the *Minimal Resolution Conjecture* (cf. [37]), require complicated computations. Therefore, it is important to have a robust (practically efficient) Computer Algebra framework, which would support this method and further experimental research (e.g. due to [40] and [41]).

Organization of the material

In this thesis we describe our computer algebra framework (implemented in SINGULAR) for computations with ideals and modules over central localizations of graded commutative algebras and give examples for its applications.

In Chapters 1 and 2 we recall some basic facts and introduce our notations.

Chapter 3 is devoted to "quotient" and "direct" Computer Algebra approaches to graded commutative algebras. Our characterization of a GB (given in Theorem 3.4.6) leads to a practically efficient algorithm for GB computation (given in Algorithm 3.4.1).

The main theoretical result of Chapter 4 is Proposition 4.3.1. It allows us to extend non-commutative computer algebra to central localizations of GR-algebras by means of rings associated to monomial orderings. In order to show that our twisted Mora Normal Form Algorithm (Algorithm 4.4.1), works over these localizations, we prove that any GR-algebra can be appropriately homogenized (Proposition 4.4.7).

Chapter 5 is devoted to computation of syzygies and free resolutions over central localizations of graded commutative algebras. We show that syzygies of leading terms can be easily computed (Proposition 5.3.4, Algorithm 5.4.1) and are of the utmost importance for GB and syzygy computations (Theorem 5.3.6, Algorithm 5.4.3). Our variation of La Scala-Stillman's Algorithm for computing Schreyer resolutions via the Schreyer frame is given in Algorithm 5.4.4.

On the implementation side we have further developed the SINGULAR non-commutative subsystem SINGULAR:PLURAL in order to allow polynomial arithmetic and more involved non-commutative basic Computer Algebra computations (e.g. S-polynomial, GB) to be easily implementable for specific algebras. At the moment graded commutative algebra-related algorithms are implemented in this framework. The developed framework is briefly described in Chapter 6.

Chapter 7 shows that our framework can be directly used for many modern interesting applications in physics and various branches of mathematics. Projective Geometry-related application is described in Section 7.1, where we give a brief overview of methods for computing sheaf cohomology. We used this application in order to test/benchmark our frameworks and further improvements. These benchmarks, listed in Section 7.1.3, show that our new algorithms and implementation are practically efficient.

In Chapter 8 we give an overview of possible further development of our framework.

Overview of our results

In contrary to the usual approach to similar quotient algebras (cf. Section 2.4) we work, define and compute GBs directly over graded commutativity algebras. Thus our algorithms are intrinsic to these algebras which enables them to be much more efficient than the general algorithms working over polynomial algebras without zero-divisors.

We also give a short account of the results by T. Stockes (cf. [123]) and prove that his definition of a Gröbner Left Ideal Basis (GLIB) is equivalent to our definition of a GB of an ideal.

We are the first to be able to compute in central localizations of non-commutative algebras (cf. Chapter 4) by allowing commutative variables to be local in general non-commutative setting and extending the Mora normal form (in the spirit of [67])

Our implementation of induced ordering (after Schreyer) provides a framework for new efficient syzygy-/free resolution- computation algorithms even in the non-commutative setting. In this framework we intend to implement a Schreyer resolution algorithm after R. La Scala and M. Stillman (cf. [81]) even over central localizations of graded commutative algebras (cf. Chapter 5).

This thesis is written with the aim of developing algorithms. We have extended and further developed the non-commutative subsystem of Singular (SINGULAR:PLURAL) into a framework, which makes it possible to seamlessly embed algebra specific efficient algorithms, as we did for our central localizations of graded commutative algebras.

Our framework makes it possible to tackle many problems more naturally and directly than it was possible before. Moreover, by implementing our approach in this framework and utilizing some commutative (linear algebra) improvements to the SINGULAR implementation of the sheaf cohomology algorithm (cf. [31]), we have achieved a speedup of approximately 10 times compared to the general implementation in SINGULAR:PLURAL, and were able to outperform M2's implementation, which was so far known to be the best one. On the other side, using these improvements and some simple (SINGULAR interpreter-related) performance tweaks, the GAP (cf. [54]) project HomAlg (cf. [9]) for general homological algebra computations (by using existing CASs for actual CA computations) achieved even better performance on our tests (while using SINGULAR as a back-end).

Similarly to HomAlg, there exists another project by D. Green and S. King, which incorporates a lot of software for the computation of cohomology rings of finite p-groups and in particular, uses our framework for all computation over graded commutative algebras (those cohomology rings are in fact graded commutative algebras). This project is briefly described in Section 7.4.

Acknowledgments

I am very grateful to my adviser Prof. G. -M. Greuel for fruitful discussions and suggestions.

Thanks go to W. Decker for his Macaulay Classic sources, to R. La Scala for an english preprint version of his Ph.D thesis, to H. Schönemann for his help and guidance to SINGULAR kernel and to V. Levandovskiyy for attracting me to non-commutative CA.

Thanks go to my parents and friends for their support. I am particularly grateful to Olena for her patience and love.

Abbreviations and Basic Notations

CA	**C**omputer **A**lgebra (GBs and related theory)
M2	**M**acaulay**2**
BBA	**B**uch**b**erger **A**lgorithm
SCA	**s**uper-**c**ommutative **a**lgebra
PBW	**P**oincaré-**B**irkhoff-**W**itt
CAS(s)	**C**omputer **A**lgebra **S**ystem(**s**)
GB, GBs	**G**röbner **B**asis, **G**röbner **B**ases
SB, SBs	**S**tandard **B**asis, **S**tandard **B**ases
iff	**if** and only **if**
w.r.t.	**w**ith **r**espect **t**o
w.l.o.g.	**w**ithout **l**oss **o**f **g**enerality
$\Bbbk$	fixed ground field
$\Bbbk^*$	the multiplicative group of units of a field $\Bbbk$
$\mathcal{A}, \mathcal{S}, \mathcal{R}, \mathcal{M}$	(non-commutative) $\Bbbk$-algebras
$\mathcal{S}^{n\|m}$ (resp. $\mathcal{A}^{n\|m}$)	Θ-algebra (resp. graded commutative algebra) with m anti-commuting variables (usually denoted by $\xi_1, \ldots, \xi_m$) and n commuting variables (usually denoted by $x_1, \ldots, x_n$)
$\mathrm{Mon}(z_1, \ldots, z_k)$	standard words (i.e. power-products or monomials) in variables $z_1, \ldots, z_k$
$\mathrm{Mon}(\mathcal{A})$	PBW basis of a PBW-algebra
$_\mathcal{A}\langle F\rangle, \langle F\rangle_\mathcal{A}, \langle F\rangle$	left, right and two-sided ideal or $\mathcal{A}$-module (resp. bimodule), generated by F
$\underline{F} = (f_1, \ldots, f_n)$	ordered tuple, for an indexed set $F = \{f_1, \ldots, f_n\}$
$\mathcal{G}_S$	the set of all finite ordered subsets of S
$\hookrightarrow, \rightarrowtail$	injective map
$\twoheadrightarrow$	surjective map

Chapter 1

Algebraic preliminaries and notations

In this chapter we recall some basic algebraic theory and introduce our notations which will be used in later chapters.

Throughout this thesis, let $\Bbbk$ denote a (commutative) field. All considered algebras and rings are associative and unital, but, in general, non-commutative. Morphisms of algebras and rings map 1 to 1.

1.1 Algebras over fields

Let us recall some notions following mostly [89] and [111, 84].

Definition 1.1.1. An *algebra over* $\Bbbk$ is an associative and unital ring $\mathcal{A}$ together with a nonzero ring homomorphism $\eta : \Bbbk \to \mathcal{A}$, which satisfy the following conditions:

- The image of η belong to the center of $\mathcal{A}$, i.e. $\eta(k) \cdot x = x \cdot \eta(k), \forall k \in \Bbbk, x \in A$.
- The map $\Bbbk \times \mathcal{A} \to \mathcal{A} : (k, x) \mapsto \eta(k) \cdot x$ turns $\mathcal{A}$ into a $\Bbbk$-vector space
- The multiplication map $\mathcal{A} \times \mathcal{A} \to \mathcal{A} : (x, y) \mapsto x \cdot y$ bi-linear, i.e. it satisfies the following conditions:

$$(x+y) \cdot z = x \cdot z + y \cdot z,$$
$$x \cdot (y+z) = x \cdot y + x \cdot z,$$
$$(\eta(k) \cdot x) \cdot y = x \cdot (\eta(k) \cdot y) = \eta(k) \cdot (x \cdot y),$$

where $k \in \Bbbk, x, y, z \in \mathcal{A}$.

Definition 1.1.2. Let $\mathcal{A}$ be an algebra over $\Bbbk$. A *subalgebra* of $\mathcal{A}$ is a $\Bbbk$-vector subspace of $\mathcal{A}$ which is closed under the multiplication of $\mathcal{A}$ and contains $1_{\mathcal{A}}$.

Remark 1.1.3. Note that due to our convention (at the beginning of the chapter) an algebras over $\Bbbk$ are associative (i.e. $x \cdot (y \cdot z) = (x \cdot y) \cdot z$ for all $x, y, z \in \mathcal{A}$) and and unital (i.e. there always exists a unique unity, that is, an element $1_\mathcal{A} \in \mathcal{A}$ such that $1_\mathcal{A} \cdot x = x \cdot 1_\mathcal{A} = x$ for all x).

In what follows unital associative algebras over the field $\Bbbk$ are called $\Bbbk$-algebras or simply algebras whenever no confusion is possible.

Remark 1.1.4. Since η (from definition 1.1.1) is nonzero it follows that η is injective, i.e. $\Bbbk \cong \operatorname{Im}(\eta)$. Thus $\Bbbk$ may be considered as a subalgebra of $\mathcal{A}$ and we may write kx instead of $\eta(k)x$ (the scalar multiplication of the $\Bbbk$-vector space $\mathcal{A}$).

Definition 1.1.5. The algebra (or ring) $\mathcal{A}$ is called ***commutative*** if $x \cdot y = y \cdot x$ for all $x, y \in \mathcal{A}$.

Definition 1.1.6. An ***algebra homomorphism*** between two $\Bbbk$-algebras is a $\Bbbk$-linear ring homomorphism.

Definition 1.1.7. An algebra over $\Bbbk$ is called ***finitely generated*** if it is finitely generated as a ring over $\Bbbk$, i.e. there exists a finitely dimensional $\Bbbk$-vector space V and a surjective $\Bbbk$-algebra homomorphism $\mathcal{T}(V) \twoheadrightarrow \mathcal{A}$, where $\mathcal{T}(V)$ is the free associative (tensor-) algebra over V (cf. 1.6.1).

Definition 1.1.8. Let $\mathcal{A}$ be an algebra over $\Bbbk$. A $\Bbbk$-subspace $I \subset \mathcal{A}$ is called a ***left ideal*** (resp., a ***right ideal***) if for any $a \in \mathcal{A}$ and $x \in I$ one has $a \cdot x \in I$ (resp., $x \cdot a \in I$).

If I is a left ideal and a right ideal, it is called a ***two-sided ideal***.

Remark 1.1.9. Let $\mathcal{A}$ be a $\Bbbk$-algebra and I be a two-sided ideal of $\mathcal{A}$. Then the factor ring (or quotient ring) $\mathcal{A}/I$ has the $\Bbbk$-algebra structure induced by $\mathcal{A}$, called the ***factor algebra*** or ***quotient algebra*** of $\mathcal{A}$. The corresponding canonical projection $\mathcal{A} \to \mathcal{A}/I$ is a $\Bbbk$-algebra homomorphism.

Definition 1.1.10. A ring (or $\Bbbk$-algebra) is called ***left-Noetherian*** if it satisfies the ascending chain condition on left ideals, that is, for any increasing sequence of left ideals $I_1 \subseteq I_2 \subseteq I_3 \subseteq \ldots$, there exists a number n such that for all $k \geqslant n$ holds $I_k = I_n$. Analogously one defines a ***right-Noetherian*** ring ($\Bbbk$-algebra).

A ring (or $\Bbbk$-algebra) is called ***Noetherian*** if it is left- and right-Noetherian.

1.2 Modules over algebras

Throughout this section let $\mathcal{A}$ be a $\Bbbk$-algebra.

Definition 1.2.1. A $\Bbbk$-vector space M, endowed with *scalar multiplication* $\cdot : \mathcal{A} \times M \to M$ is called a ***left $\mathcal{A}$-module*** if the following conditions are satisfied for all $a, b \in \mathcal{A}$ and

1.2. MODULES OVER ALGEBRAS

$m, n \in M$:

$$1_{\mathcal{A}} \cdot m = m,$$
$$(ab) \cdot m = a \cdot (b \cdot m),$$
$$(a+b) \cdot m = (a \cdot m) + (b \cdot m),$$
$$a \cdot (m+n) = (a \cdot m) + (a \cdot n).$$

Analogously to Definition 1.2.1 we can define a right module. A left module which is also a right module is called a **bimodule**. From now on we will say simply "module" instead of "left module".

Remark 1.2.2. One-sided ideals in algebras (rings) are particular cases of modules over these algebras (rings), e.g. any $\Bbbk$-algebra can be considered as a module over itself.

Definition 1.2.3. Let M be an $\mathcal{A}$-module. A $\Bbbk$-subvector space $N \subset M$ is called an $\mathcal{A}$-**submodule** or just a **submodule** of M if $a \cdot m \in N$, and $m + n \in N$ for all $a \in \mathcal{A}, m, n \in N$.

Definition 1.2.4. If M and N are (left) $\mathcal{A}$-modules, then a $\Bbbk$-linear map $\psi : M \to N$ is an **homomorphism of $\mathcal{A}$-modules** if $\psi(a \cdot x + b \cdot y) = a \cdot \psi(x) + b \cdot \psi(y)$, for all $x, y \in M$ and $a, b \in \mathcal{A}$.

A bijective module homomorphism is an *isomorphism of modules*, and the two modules are called *isomorphic*.

The isomorphism theorems familiar from groups and vector spaces are also valid for $\mathcal{A}$-modules.

Definition 1.2.5. Let M be an $\mathcal{A}$-module and $N \subseteq M$ be a submodule. We say that $m_1, m_2 \in M$ are equivalent w.r.t. N if $m_1 - m_2 \in N$. Denote the equivalence class of $m \in M$ w.r.t. N by $m + N$ or $[m]$. Each element in the class $m + N$ is called a representative of the class.

The *factor module* or *quotient module* is $M/N := \{m + N \mid m \in M\}$, with the operations induced by operations on representatives.

Remark 1.2.6. Let ψ be an $\mathcal{A}$-module homomorphism from M to N. Denote its **kernel** by

$$\mathrm{Ker}(\psi) := \{m \in M \mid \psi(m) = 0\} \subseteq M,$$

its **image** by

$$\mathrm{Im}(\psi) := \{\psi(m) \mid m \in M\} \subseteq N,$$

and its **cokernel** by

$$\mathrm{Coker}(\psi) := N/\mathrm{Im}(\psi).$$

Trivial verification shows that $\mathrm{Ker}(\psi)$ and $\mathrm{Im}(\psi)$ are $\mathcal{A}$-submodules in respectively M and N. Hence the definition of cokernel is correct and, obviously, $\mathrm{Coker}(\psi)$ is an $\mathcal{A}$-module.

Definition 1.2.7. Let M be a left $\mathcal{A}$-module, $S \subset M$. We denote by ${}_{\mathcal{A}}\langle S \rangle$ the left submodule in M generated by S over $\mathcal{A}$:

$$ {}_{\mathcal{A}}\langle S \rangle := \left\{ \sum_{finite} a_m \cdot m \mid m \in S, a_m \in \mathcal{A} \right\}. $$

We will write ${}_{\mathcal{A}}\langle m_1, \ldots, m_k \rangle$ instead of ${}_{\mathcal{A}}\langle \{m_1, \ldots, m_k\} \rangle$.

Definition 1.2.8. A left $\mathcal{A}$-module M is called **finitely generated** if there exist finitely many vectors $m_1, \ldots, m_k \in M$ such that $M = {}_{\mathcal{A}}\langle m_1, \ldots, m_k \rangle$.

Remark 1.2.9. Due to remark 1.2.2 we can use the same notation for left ideals generated by subsets in $\mathcal{A}$.

Definition 1.2.10. Let M be a left $\mathcal{A}$-module. The **annulator** of M is the following left ideal in $\mathcal{A}$:
$$ \mathrm{Ann}(M) := \{ a \in \mathcal{A} \mid a \cdot m = 0 \, \forall m \in M \}. $$

For any $\mathcal{A}$-module element m, by abuse of notation, we define:
$$ \mathrm{Ann}(m) := \{ a \in \mathcal{A} \mid a \cdot m = 0 \}. $$

1.3 Noetherian modules

Definition 1.3.1. A (left) $\mathcal{A}$-module M is called (left-) **Noetherian** whenever it satisfies any of the following conditions.

1. Every (left) submodule of M is finitely generated.
2. It satisfies ascending chain condition on (left) submodules, i.e. every ascending sequence of submodules of M: $M_1 \subset M_2 \subset \ldots$ such that $M_i \neq M_{i+1}$ is finite.
3. Every non-empty set of (left) submodules of M has a maximal element.

The proof that the three conditions from definition 1.3.1 are equivalent, and for the following propositions can be found in [84, Chapter VI].

Proposition 1.3.2. Let M be a (left) Noetherian $\mathcal{A}$-module. Then every submodule and every factor module of M is Noetherian.

Proposition 1.3.3. Let M be an $\mathcal{A}$-module and N be a submodule. If N and M/N are Noetherian then M is Noetherian.

Thus, an exact sequence of (left) $\mathcal{A}$-modules $0 \to M' \to M \to M'' \to 0$, M is Noetherian iff both M', M'' are Noetherian.

1.4. FREE MODULES OVER ALGEBRAS

Corollary 1.3.4. *Let M be an $\mathcal{A}$-module, N and N' be submodules. If $M = N + N'$ and if both N and N' are Noetherian then M is Noetherian. Thus the finite direct sum of Noetherian modules is Noetherian.*

It's easy to see that a ring (or $\Bbbk$-algebra) $\mathcal{A}$ is Noetherian iff it is Noetherian as a left module over itself. It follows that every left ideal in $\mathcal{A}$ is finitely generated.

Proposition 1.3.5. *Let $\mathcal{A}$ be a (left-)Noetherian ring (or $\Bbbk$-algebra) and let M be a finitely generated (left-)$\mathcal{A}$-module. Then M is (left-)Noetherian.*

Proposition 1.3.6. *Let $\mathcal{A}$ be a Noetherian ring (or a $\Bbbk$-algebra) and let $\psi : \mathcal{A} \to \mathcal{B}$ be a surjective ring (or $\Bbbk$-algebra) homomorphism. Then $\mathcal{B}$ is Noetherian.*

1.4 Free modules over algebras

In what follows let $\mathcal{A}$ be a Noetherian $\Bbbk$-algebra. We denote by $\mathcal{A}^r$ the free left $\mathcal{A}$-module of rank r, i.e. direct product of r copies of $\mathcal{A}$, with scalar multiplication defined component-wise. Let $\epsilon_i := (0, \ldots, 0, 1, 0, \ldots, 0)^t \in \mathcal{A}^r$, where 1 is in the i-th component. The vectors $\epsilon_1, \ldots, \epsilon_r$ form the so-called **canonical (free) basis** for $\mathcal{A}^n$. Which means, in particular, that any $v = (v_1, \ldots, v_r)^t \in \mathcal{A}^n$ can be uniquely written as $v = \sum_{i=1}^r v_i \cdot \epsilon_i$.

Remark 1.4.1. Let M be a left $\mathcal{A}$-module and $\underline{m} = (m_1, \ldots, m_k) \in M \times \ldots \times M$. Consider the associated $\mathcal{A}$-module homomorphism (left $\mathcal{A}$-liner map), uniquely defined by

$$\psi_{\underline{m}} : \mathcal{A}^k \to M : \epsilon_i \mapsto m_i.$$

Clearly the homomorphism is surjective iff the elements m_i generate M, in which case

$$M \cong \mathcal{A}^k / \operatorname{Ker}(\psi_{\underline{m}}). \tag{1.1}$$

Conversely, if $N \subset \mathcal{A}^k$ is any submodule, then $\mathcal{A}^k/N$ is a finitely generated $\mathcal{A}$-module. Therefore one can view finitely generated (left) $\mathcal{A}$-modules (together with a finite set of generators) as being essentially the same thing as quotients of free (left) $\mathcal{A}$-modules of finite rank.

We will sometimes use the following notation for the image of $S = \sum_{i=1}^k s_i \epsilon_i \in \mathcal{A}^k$ under the map defined by $\underline{m} \in M^k$:

$$S * \underline{m} := \psi_{\underline{m}}(S) = \sum_{i=1}^k s_i \cdot m_i$$

Definition 1.4.2. A **syzygy** or **relation** between k elements $m_1, \ldots, m_k \in M$ is a vector $S = \sum_{i=1}^{k} s_i \epsilon_i \in \mathcal{A}^k$ such that

$$S * (m_1, \ldots, m_k) = 0.$$

The set of all syzygies between $m_1, \ldots, m_k$ is a (left) submodule in $\mathcal{A}^k$. Indeed, it is the kernel of the map $\psi_{\underline{m}}$. We denote it by $\mathrm{Syz}(m_1, \ldots, m_k)$ and call it the **syzygy module** of $m_1, \ldots, m_k$.

Note that the syzygy module of any finitely generated $\mathcal{A}$-module is finitely generated since $\mathcal{A}$ is Noetherian.

If the elements m_i generate M then there exists the exact sequence

$$\mathcal{A}^l \xrightarrow{\Phi} \mathcal{A}^k \xrightarrow{\psi_{\underline{m}}} M \to 0, \tag{1.2}$$

where the exactness means that $\mathrm{Im}(\Phi) = \mathrm{Ker}(\psi_{\underline{m}}) = \mathrm{Syz}(m_1, \ldots, m_k)$. Therefore due to equation (1.1)

$$M \cong \mathcal{A}^k / \mathrm{Im}(\Phi) = \mathrm{Coker}(\Phi)$$

So that we can represent the module M as the cokernel of the following map defined by the matrix Φ:

$$\mathcal{A}^l \ni (w_1, \ldots, w_l)^t \mapsto \left((w_1, \ldots, w_l) \cdot \Phi^t\right)^t \in \mathcal{A}^k. \tag{1.3}$$

Definition 1.4.3. For a finitely generated $\mathcal{A}$-module M, a (possibly infinite) exact sequence [1]

$$\cdots \to \mathbf{F}_{i+1} \xrightarrow{\varphi_{i+1}} \mathbf{F}_i \to \cdots \to \mathbf{F}_1 \xrightarrow{\varphi_1} \mathbf{F}_0 \xrightarrow{\varphi_0} M \to 0,$$

with finitely generated free $\mathcal{A}$-modules F_i is called a **free resolution** of M by left $\mathcal{A}$-modules, where the maps φ_{i+1} are given by some matrices as in (1.3).

Note that there always exists a free resolution. One, for instance, can be constructed by extending the short exact sequence from (1.2).

1.5 Graded structures

Definition 1.5.1. Let Γ be a commutative additive monoid (usually an Abelian group), whose identity element is denoted by 0. A **Γ-graded ring** $\mathcal{A}$ is a ring together with a

---
[1] Sometimes the complex of free $\mathcal{A}$-modules

$$\mathbf{F}_\bullet : \cdots \to \mathbf{F}_{i+1} \xrightarrow{\varphi_{i+1}} \mathbf{F}_i \to \cdots \to \mathbf{F}_1 \xrightarrow{\varphi_1} \mathbf{F}_0,$$

with $\mathrm{Coker}(\varphi_1) \cong M$ is called a free resolution of M.

1.5. GRADED STRUCTURES

direct sum decomposition into (Abelian) additive subgroups of the additive subgroup of $\mathcal{A}$:

$$\mathcal{A} = \bigoplus_{g \in \Gamma} \mathcal{A}_g,$$

such that $\mathcal{A}_i * \mathcal{A}_j \subseteq \mathcal{A}_{i+j}$, for all $i, j \in \Gamma$. For $d \in \Gamma$, $\mathcal{A}_d$ is called the **homogeneous part of degree d** of $\mathcal{A}$. Elements from $\mathcal{A}_d$ are called **homogeneous elements of degree d** and denote the degree of any such element a by $|a| := d$. A k-algebra $\mathcal{A}$ is Γ-graded whenever it is Γ-graded as a ring and all homogeneous parts are k-subvector spaces of the k-vector space $\mathcal{A}$.

Example 1.5.2. Let us give some examples:

- Any commutative polynomial k-algebra $k[x_1, \ldots, x_n]$ is $\mathbb{Z}$-graded by degree.
- A $\mathbb{Z}_2$-graded k-algebra is also called a **super-algebra** (see also Section 7.3).
- Clearly if there is a surjective homomorphism $\Gamma \to \Gamma'$ then any Γ-graded algebra can be considered to be Γ'-graded. In particular any ($\mathbb{Z}$-)graded commutative algebra (cf. Section 1.6) can be endowed with a $\mathbb{Z}_2$-grading if we take the degree modulo 2 (e.g. $|\xi_1| = |\xi_1 \xi_2 \xi_3| \equiv \underline{1} \pmod{2}$, and $|x_1| = |\xi_1 \xi_2| \equiv \underline{0} \pmod{2}$, by putting $|x_j| := \underline{0}$ for commutative variables and $|\xi_i| := \underline{1}$ for anti-commutative variables). This way any graded commutative algebra can be considered as a **super-commutative algebra** (cf. Remark 1.5.11).

Remark 1.5.3. Let $\mathcal{A}$ be a Γ-graded ring (resp. k-algebra). Then directly from Definition 1.5.1 it follows that $\mathcal{A}_0$ is a subring (resp. subalgebra) of $\mathcal{A}$, and $1_\mathcal{A} \in \mathcal{A}_0$ is an element of degree zero.

Definition 1.5.4. Let $\mathcal{A}$ be a Γ-graded ring. An $\mathcal{A}$-module M is called $\boldsymbol{\Gamma}$**-graded** if

$$M = \oplus_{g \in \Gamma} M_g,$$

where the M_g are subgroups of the additive group of M, such that $\mathcal{A}_i * M_j \subset M_{i+j}$, for all $i, j \in \Gamma$. For $d \in \Gamma$, M_d is called the **homogeneous part of degree d** of M. Elements from M_d are called **homogeneous elements of degree d**. Whenever $\mathcal{A}$ is a k-algebra it is additionally required that all homogeneous parts M_g are k-subvector spaces of k-vector space M.

Remark 1.5.5. Clearly, for any $a \in \mathcal{A} = \oplus_{g \in \Gamma}$ there is a unique (finite) decomposition of a into homogeneous parts: $a = \sum_{g \in \Gamma} a_g$, with $a_g \in \mathcal{A}_g$.

Analogously, if M is a Γ-graded $\mathcal{A}$-module, every $m \in M$ has a unique (finite) decomposition of m into homogeneous parts: $m = \sum_{g \in \Gamma} m_g$, with $m_g \in M_g$.

Proposition 1.5.6 (Proposition 3.3 [89]). *Let $\mathcal{A}$ be a Γ-graded ring, M be a graded $\mathcal{A}$-module and N be a submodule of M. Then the following conditions are equivalent:*

(i) $N = \oplus_{g \in \Gamma} (M_g \cap N)$,

(ii) for any $u \in N$ its homogeneous parts u_g belong to N,
(iii) N is generated by homogeneous elements,
(iv) the factor $\mathcal{A}$-module $M/N = \sum_{g \in \Gamma}(M_g + N)/N$ is Γ-graded with the decomposition

$$M/N = \bigoplus_{g \in \Gamma}(M_g + N)/N.$$

Definition 1.5.7. Let $\mathcal{A}$ be a Γ-graded ring and M be a graded $\mathcal{A}$-module. A submodule $N \subset M$ satisfying any of the equivalent conditions of Proposition 1.5.6 is called *graded submodule* of M.

In particular, since left (two-sided) ideals of $\mathcal{A}$ are submodules, we get the definition of *graded left (two-sided) ideal* by replacing submodules by (left or two-sided) ideals in the above.

Remark 1.5.8. For a finitely generated graded $\mathcal{A}$-module $M = \bigoplus_{i \in \mathbb{Z}} M_i$, we shall denote the twisted by degree d $\mathcal{A}$-module by

$$M(d) := \bigoplus_{i \in \mathbb{Z}} M_{i+d},$$

and the truncated at degree d $\mathcal{A}$-module by

$$M_{\geqslant d} := \bigoplus_{i \in \mathbb{N}} M_{i+d}.$$

Definition 1.5.9. Let $\mathcal{A}$ and $\mathcal{B}$ be two Γ-graded rings (resp. algebras). A ring (resp. algebra) homomorphism $\psi : \mathcal{A} \to \mathcal{B}$ is called a *graded ring (resp. algebra) homomorphism (of degree 0)* if $\psi(\mathcal{A}_g) \subset \mathcal{B}_g$ for all $g \in \Gamma$.

Let M and N be two graded modules over a Γ-graded ring (algebra) $\mathcal{A}$. An $\mathcal{A}$-homomorphism $\psi : M \to N$ is called a *graded $\mathcal{A}$-homomorphism (of degree 0)* if $\psi(M_g) \subset N_g$ for all $g \in \Gamma$.

As a simple corollary of the above definitions and Proposition 1.5.6, we get that images and kernels of graded ring (algebra) homomorphisms (resp. graded $\mathcal{A}$-homomorphisms) are graded subrings (subalgebras) (resp, submodules).

Definition 1.5.10 (following [100]). Let Γ be an additive Abelian group endowed with a bilinear map $\langle,\rangle : \Gamma \times \Gamma \to \mathbb{Z}_2$. A Γ-graded ring (resp. algebra) $\mathcal{A}$ is called *Γ-graded commutative* (or *Γ-commutative*) if $a * b = (-1)^{\langle |a|, |b| \rangle} b * a$ holds for all homogeneous $a, b \in \mathcal{A}$.

Remark 1.5.11. Note that a Γ-graded commutative $\Bbbk$-algebra is also called *graded commutative* (resp. *super-commutative*) in the case when Γ is $\mathbb{Z}$ (resp. $\mathbb{Z}_2$) and the bilinear map is the multiplication.

1.6 Tensor algebra

Let V be a vector space over $\Bbbk$. The tensor algebra of V, denoted $\mathcal{T}(V)$, is the algebra of tensors on V (of any rank) with multiplication being the tensor product. The tensor algebra is, in a sense, the "most general" algebra containing V. This is formally expressed by a certain universal property (see below).

Definition 1.6.1. The ***tensor algebra over V***, denoted by $\mathcal{T}(V)$ is the $\mathbb{Z}$-graded $\Bbbk$-algebra with the n-th graded component given by n-th tensor power of V:

$$\mathcal{T}_n(V) := V^{\otimes n} = \overbrace{V \otimes \cdots \otimes V}^{n \text{ times}}, \quad n = 1, 2, \ldots,$$

and $\mathcal{T}_0(V) := \Bbbk$. That is,

$$\mathcal{T}(V) := \bigoplus_{n=0}^{\infty} \mathcal{T}_n(V).$$

The multiplication $m : \mathcal{T}(V) \times \mathcal{T}(V) \to \mathcal{T}(V)$ is determined by the canonical isomorphism $\mathcal{T}_k(V) \otimes \mathcal{T}_l(V) \to \mathcal{T}_{k+l}(V)$ given by the tensor product:

$$m(a, b) = a \otimes b, \quad a \in V^{\otimes k}, \, b \in V^{\otimes l},$$

which is then extended by linearity to all of $\mathcal{T}(V)$.

Remark 1.6.2. The construction generalizes in straightforward manner to the tensor algebra of any module M over a commutative ring R. If R is a non-commutative ring, one can still perform the construction for any R-R bimodule M. This does not work for ordinary R-modules because the iterated tensor products cannot be formed.

The fact that the tensor algebra is the most general algebra containing V is expressed by the following **universal property**:

Any linear transformation $f : V \to A$ from V to a $\Bbbk$-algebra A can be uniquely extended to an algebra homomorphism from $\mathcal{T}(V)$ to A as indicated by the following commutative diagram:

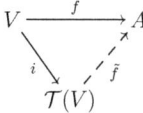

Here i is the canonical inclusion of V into $\mathcal{T}(V)$, identifying it with $\mathcal{T}_1(V)$. In fact, one can define the tensor algebra $\mathcal{T}(V)$ as the unique $\Bbbk$-algebra satisfying this universal property (moreover, $\mathcal{T}(V)$ is unique up to a unique isomorphism).

The tensor algebra $\mathcal{T}(V)$ is also called the ***free associative algebra*** on the vector space V.

If V has finite dimension n, its tensor algebra can be regarded as the "algebra of polynomials over $\Bbbk$ in n non-commuting variables". If we take basis vectors for V, those become non-commuting variables in $\mathcal{T}(V)$, subject to no constraints (beyond associativity, the distributive law and $\Bbbk$-linearity). That is, we construct the free associative algebra of V in the following way: choose a basis $B = \{x_i \mid i \in I\}$ in V, and let

$$T := \Bbbk \langle B \rangle := \Bbbk \langle x_i \mid i \in I \rangle$$

be the algebra of non-commutative polynomials in variables $\{x_i\}$ with coefficients in $\Bbbk$. As a vector space, it is generated by monomials in these variables which are finite sequences of x_i in arbitrary order (repetitions are allowed). The product is defined by concatenation of the monomials. The map $i : V \to T$ is the natural embedding $(x_i \mapsto x_i)$. This construction of a free $\Bbbk$-algebra on the set B as an algebra of words is given in [89, Chapter 1].

Definition 1.6.3. The finitely generated (by ordered symbols $x_1, \ldots, x_n$) free associative $\Bbbk$-algebra $\Bbbk \langle x_1, \ldots, x_n \rangle$ is called ***general non-commutative polynomial ring*** over the field $\Bbbk$. Its basis consists of power products $x_{i_1}^{\alpha_1} x_{i_2}^{\alpha_2} \cdots x_{i_m}^{\alpha_m}$, called ***words*** where $m \geqslant 0, \alpha_i \geqslant 0$ and $1 \leqslant i_1, i_2, \ldots, i_m \leqslant n$.

Power products with ordered indices $1 \leqslant i_1 < i_2 < \ldots < i_m \leqslant n$ shall be called ***standard words*** Their finite linear combinations will be called ***standard polynomials***. Let us denote the set of all standard words in symbols $x_1, \ldots, x_n$ by $\mathrm{Mon}(x_1, \ldots, x_n)$.

Remark 1.6.4. Due to Definition 1.6.3, every nonzero standard polynomial f has a unique representation, as a sum of standard words with nonzero coefficients:

$$f = \sum_{\mathbf{x}^{\underline{\alpha}} \in \mathrm{Mon}(x_1, \ldots, x_n)} c_{\underline{\alpha}} \mathbf{x}^{\underline{\alpha}}.$$

We define the ***support*** of f by

$$\mathrm{Supp}(f) := \{\underline{\alpha} \in \mathbb{N}^n : c_{\underline{\alpha}} \neq 0\} \subseteq \mathbb{N}^n,$$

and the corresponding set of standard words by

$$\mathrm{Mon}(f) := \{\mathbf{x}^{\underline{\alpha}} : c_{\underline{\alpha}} \neq 0\} \subseteq \mathrm{Mon}(x_1, \ldots, x_n),$$

Definition 1.6.5. Let $\preccurlyeq$ be a fixed total order on $\mathbb{N}^n$ (cf. the coming section 2.1). The ***leading exponent*** of a nonzero standard polynomial f is defined by

$$\mathrm{Exp}(f) := \max_{\preccurlyeq} \mathrm{Supp}(f).$$

Definition 1.6.6. Let $\preccurlyeq$ be a fixed total order on $\mathrm{Mon}(x_1, \ldots, x_n)$. The ***leading monomial*** of a nonzero standard polynomial f is defined by

$$\mathrm{Lm}(f) := \max_{\preccurlyeq} \mathrm{Mon}(f).$$

The corresponding (nonzero) coefficient is denoted by $\mathrm{Lc}(f)$ and is called the ***leading coefficient***. The ***leading term*** of f is defined by $\mathrm{Lt}(f) := \mathrm{Lc}(f) \cdot \mathrm{Lm}(f)$. The ***tail*** of f is defined by $\mathrm{Tail}(f) := f - \mathrm{Lt}(f)$.

1.6. TENSOR ALGEBRA

Because of the generality of the tensor algebra, all other algebras are constructed by starting with the tensor algebra and then imposing certain relations on the generators, that is, by constructing certain factors of $\mathcal{T}(V)$.

Proposition 1.6.7 (cf. [89], Prop. 1.2). *Any $\Bbbk$-algebra $\mathcal{A}$ is isomorphic to a factor of a free associative $\Bbbk$-algebra by some two-sided ideal.*

Definition 1.6.8. Let $B = \{x_j\}$ be a set of symbols. Let I be the two-sided ideal in $\Bbbk\langle B\rangle$ generated by $\mathcal{F} := \{f_i\} \subset \Bbbk\langle B\rangle$. We say that the $\Bbbk$-algebra $\mathcal{A} = \Bbbk\langle B\rangle/I$ is ***generated by B subject to relations $\mathcal{F}$*** and denoted by $\Bbbk\langle x_j \mid f_i = 0\rangle$. The relations f_i are called ***defining relations*** for $\mathcal{A}$.

Definition 1.6.9. Let $\mathcal{A}$ be a $\Bbbk$-algebra generated by a finite (ordered) set of symbols $\{x_1, \ldots, x_n\}$ subject to some relations. The algebra $\mathcal{A}$ is said to be a ***Poincaré-Birkhoff-Witt-algebra*** (or simply ***PBW-algebra***), if the following set of standard words is a generating system of $\mathcal{A}$ as a $\Bbbk$-vector space:

$$\{\mathbf{x}^{\underline{\alpha}} := x_1^{\alpha_1} x_2^{\alpha_2} \cdots x_n^{\alpha_n} \mid \alpha_i \geqslant 0\}$$

Remark 1.6.10. Due to definition 1.6.9, we can choose a suitable $\Bbbk$-vector space basis of $\mathcal{A}$ consisting of standard words. Let us consider PBW-algebras together with a such basis, called a ***PBW-basis*** and denoted by $\mathrm{Mon}(\mathcal{A})$.

Moreover, after fixing a total order on $\mathrm{Mon}(\mathcal{A})$ we can use the notions of leading monomial, coefficient and term as they where introduced for a free associative algebra (cf. definition 1.6.6).

Clearly, the free associative non-commutative polynomial algebra $\Bbbk\langle x_1, \ldots, x_n\rangle$ does not have a PBW basis.

Examples 1.6.11. Let us give defining relations for the most important algebras dealt with within this thesis. Note that they all have PBW bases.

The ***symmetric algebra in n variables*** (also called commutative polynomial algebra), denoted by $\Bbbk[x_1, \ldots, x_n]$), is defined as follows:

$$\Bbbk\langle x_1, \ldots, x_n \mid x_i \cdot x_j - x_j \cdot x_i = 0\rangle.$$

Its PBW-basis shall consist of all power-products in all the variables $x_1, \ldots, x_n$.

Note that every commutative algebra is a quotient of a symmetric algebra.

The ***exterior algebra in m variables over a commutative $\Bbbk$-algebra A*** is defined as the $\Bbbk$-algebra generated over A by elements $\xi_1, \ldots, \xi_m$ subject to the relations

$$\xi_i^2 = 0, \ \xi_i \cdot \xi_j = -\xi_j \cdot \xi_i, i \neq j,$$

and in our notation above, will be denoted by

$$\mathcal{A} := A\langle \xi_1, \ldots, \xi_m \mid \xi_i^2 = \xi_i \cdot \xi_j + \xi_j \cdot \xi_i = 0, i \neq j \rangle.$$

Its *PBW*-basis shall consist of square-free power-products in all the variables $\xi_1, \ldots, \xi_m$.

It is clear that this algebra can be constructed as the tensor product of A and the usual exterior algebra:

$$\mathcal{A} \cong A \otimes_{\Bbbk} \Bbbk\langle \xi_1, \ldots, \xi_m \mid \xi_i^2 = \xi_i \cdot \xi_j + \xi_j \cdot \xi_i = 0, i \neq j \rangle.$$

An exterior algebra over the commutative polynomial algebra $\Bbbk[x_1, \ldots, x_n]$ is one of the most important objects for us. It may be constructed as follows:

$$\Bbbk\langle x_1, \ldots, x_n, \xi_1, \ldots, \xi_m \mid x_i \cdot x_j - x_j \cdot x_i = x_i \cdot \xi_l - \xi_l \cdot x_i = \xi_k \cdot \xi_l + \xi_l \cdot \xi_k = 0 \rangle.$$

In this thesis such an algebra will be called a *(polynomial) graded commutative algebra* in n commuting (or commutative or even) variables $x_1, \ldots, x_n$ and m anti-commuting (or anti-commutative or odd) variables $\xi_1, \ldots, \xi_m$.

Its *PBW*-basis shall consist of those power-products, which do not contain powers of variables $\xi_1, \ldots, \xi_m$.

Note that this algebra is $\mathbb{Z}$-graded commutative and even super-commutative in the sense of Remark 1.5.11, if we set odd degrees for commutative and even degrees for anti-commuting variables.

1.7 Binary relations

Definition 1.7.1. A *binary relation* R on a set S is a subset R of $S \times S$. We say that "a is R-related to b" if $(a, b) \in R$ and denote it by aRb.

A binary relation R on a set S is called

- *reflexive* if $\forall a \in S : aRa$. For example: "is greater than or equal to" but not "greater than".
- *irreflexive* if $\forall a \in S : (a, a) \notin R$. For example: "greater than" but not "greater than or equal to".
- *symmetric* if $\forall a, b \in S$ it holds that if aRb then bRa. For example: "is a blood relative of".
- *antisymmetric* if $\forall a, b \in S$ it holds that if aRb and bRa then $a = b$. For example: "is greater than or equal to".
- *asymmetric* if $\forall a, b \in S$ it holds that if aRb then bRa is false. For example: "is greater than".

1.7. BINARY RELATIONS

- **transitive** if $\forall a, b, c \in S$ it holds that if aRb and bRc then aRc. For example: "is an ancestor of".
- **total** if $\forall a, b \in S$ it holds that either aRb or bRa is true. Such a relation is also called **linear order** For example: "is greater than or equal to" on $\mathbb{Z}$.
- an **equivalence relation** if R is reflexive, symmetric and transitive. For example: "equals to".
- a **quasi-order** if it is irreflexive and transitive.
- a **partial order** if it is reflexive, antisymmetric and transitive.
- a **total order** if it is a partial order and total.
- a **well-order** if it is a total order w.r.t. which every nonempty subset $B \subset S$ has a least element, where an element $c \in B$ is called the **least element** of B if cRb for every $b \in B$. In such a case the set S is said to be **well ordered**

Let us consider some operations on binary relations:

- **Reflexive closure** of R is defined by $R^= := R \cup \{(a, a) \mid a \in S\}$.
- **Transitive closure** of R, denoted by R^+, is the smallest transitive relation over S containing R or equivalently the intersection of all transitive relations containing R.
- The reflexive transitive closure of R is defined by $R^* := (R^+)^=$.
- **Symmetric closure** of R is defined by $R^- := R \cup \{(b, a) \mid (a, b) \in R\}$.
- The reflexive symmetric transitive closure of R shell be called the *equivalence relation defined by* R and denoted by $\overleftrightarrow{R}$.

An element $a \in S$ is a **normal form** w.r.t. R (or modulo R) if there is no $b \in S : (a, b) \in R$. Such an elements may also be called *minimal* element of S w.r.t. the binary relation R.

The binary relation R is Noetherian if there is no infinite sequence $\{f_i\}_{i \in \mathbb{N}} \subset S$ such that $(f_i, f_{i+1}) \in R$ for all $i \in \mathbb{N}$.

The following is due to [123] and references thereof.

Definition 1.7.2 (Definitions. 5.7 − 5.9). Let R be a binary relation on a set S. Let closures R^*, $\overleftrightarrow{R}$ be as above. The binary relation R is called

- **confluent** if for all $f, a, b \in S : fR^*a$ and fR^*b there exists $g \in S : aR^*g$ and bR^*g.
- **locally confluent** if for all $f, a, b \in S : fRa$ and fRb there exists $g \in S : aR^*g$ and bR^*g.
- **Church–Rosser** if for all $a, b \in S : a\overleftrightarrow{R}b$ there exists $f \in S : aR^*f$ and bR^*f.

Lemma 1.7.3 (Lemma 5.10). *Let R be a Noetherian binary relation on a set S. Then the following statements are equivalent*

- *R is confluent.*

- R is locally confluent.
- R is Church–Rosser.
- For all $a, b \in S$: $a \overset{\leftrightarrow}{R} b$ iff there exists a normal form $f \in S$ w.r.t. R such that aR^*f and bR^*f.

Chapter 2

Computer Algebra preliminaries

In this chapter we recall CA basis for (polynomial) G-algebras and the possible approaches to CA computations over quotient algebras (e.g. GR-algebras or graded commutative algebras).

2.1 Standard monomials and monomial orderings

In this section we will discuss monoid and monomial orderings mainly following [19, Chapter 2].

Definition 2.1.1. Let $\preccurlyeq$ be a partial ordering, i.e., a reflexive, antisymmetric and transitive relation, on a non-empty set M.

A partial ordering $\preccurlyeq$ on a set M satisfies the ***descending chain condition*** if there exists no infinite strictly descending chain
$$\gamma_1 \succ \gamma_2 \succ \ldots \succ \gamma_n \succ \ldots, \quad \gamma_i \in M.$$

Proposition 2.1.2. *A partial ordering $\preccurlyeq$ on a set M satisfies the descending chain condition iff every non-empty subset of M has a minimal element.*

Recall that a (multiplicative) monoid $(M, \cdot)$ with neutral element $e \in M$ is a set M endowed with a binary operation $\cdot$ which is associative and the neutral element satisfies $e \cdot m = m \cdot e = m$ for all $m \in M$.

Example 2.1.3. Let X be a non-empty set, called alphabet. A word or a term over X is an ordered finite sequence $x_1 \cdots x_s$ of elements $x_i \in X$. Adding the empty sequence, denoted by 1, to the set of words over X, we obtain the ***free monoid*** on X, denoted by $\langle X \rangle$. The multiplication in $\langle X \rangle$ is just the concatenation of words and 1 acts as neutral element. The characteristic property of X is that it is a free object, i.e., any mapping $X \to M$, where M is a monoid, extends uniquely to a *homomorphism of monoids* $\langle X \rangle \to M$, that is, a multiplicative map $\langle X \rangle \to M$, which maps the neutral element 1 of $\langle X \rangle$ in to the neutral element of M.

We are especially interested in the case when X is finite, say $X = \{x_1, \ldots, x_n\}$. In this case we use the notation $\langle X \rangle = \langle x_1, \ldots, x_n \rangle$.

Example 2.1.4. Let n be a positive integer and let

$$\mathbb{N}^n = \{\underline{\alpha} = (\alpha_1, \ldots, \alpha_n) : \alpha_1, \ldots, \alpha_n \in \mathbb{N}\}.$$

We will consider the commutative monoid $(\mathbb{N}^n, +)$ with sum defined component-wise. The neutral element is then given by $\underline{0} = (0, \ldots, 0)$.

Definition 2.1.5. Let $(M, \cdot)$ be a monoid. A total or partial ordering $\prec$ on M is called **monoid ordering** if

$$\forall m_1, m_2, a, b \in M : (m_1 \prec m_2) \text{ or } (m_1 = m_2) \Rightarrow (am_1 b \prec am_2 b) \text{ or } (am_1 b = am_2 b). \tag{2.1}$$

Remark 2.1.6. If the monoid M is **cancelative** (i.e. when either $am = bm$ or $ma = mb$ implies $a = b$) then condition (2.1) in definition 2.1.5 can be replaced by

$$\forall m_1, m_2, a, b \in M : m_1 \prec m_2 \Rightarrow am_1 b \prec am_2 b. \tag{2.2}$$

Example 2.1.7. Clearly monoids $\langle X \rangle$ and $(\mathbb{N}^n, +)$ (cf. examples 2.1.3 and 2.1.4) are cancelative, in particular, since $(\mathbb{N}^n, +)$ is a commutative monoid: monomial ordering on $(\mathbb{N}^n, +)$ is a partial monoid ordering $\prec$ such that $\underline{\alpha} + \underline{\gamma} \prec \underline{\beta} + \underline{\gamma}$ for all $\underline{\alpha}, \underline{\beta}, \underline{\gamma} \in \mathbb{N}^n$ with $\underline{\alpha} \prec \underline{\beta}$.

Definition 2.1.8. A non-empty subset E of $\mathbb{N}^n$ is said to be a **monoideal** if $E + \mathbb{N}^n = E$. If B is a subset of $\mathbb{N}^n$, then we define the *monoideal* generated by B to be

$$B + \mathbb{N}^n = \bigcup_{\underline{\beta} \in B} (\underline{\beta} + \mathbb{N}^n) = \{\underline{\beta} + \underline{\gamma}; \underline{\beta} \in B, \underline{\gamma} \in \mathbb{N}^n\}.$$

If $E = B + \mathbb{N}^n$, then we call the elements of B *generators* of E.

Definition 2.1.9. The (natural) partial ordering $\preccurlyeq^n$ in $\mathbb{N}^n$ is defined by

$$\underline{\alpha} \preccurlyeq^n \underline{\beta} \Leftrightarrow \underline{\beta} \in \underline{\alpha} + \mathbb{N}^n.$$

In other words, $\underline{\alpha} \preccurlyeq^n \underline{\beta}$ if $\alpha_i \leqslant \beta_i$ for all $1 \leqslant i \leqslant n$.

This natural partial ordering may sometimes be denoted by $\leqslant_{nat}$.

Clearly, the partial ordering $\preccurlyeq^n$ satisfies the descending chain condition.

2.1. STANDARD MONOMIALS AND MONOMIAL ORDERINGS

Lemma 2.1.10 (Dickson's Lemma, cf. [33]). *For any non-empty $E \subseteq \mathbb{N}^n$, there exists a finite subset $B = \{\underline{\alpha_1}, \ldots, \underline{\alpha_m}\}$ of E ($\underline{\alpha_m}$ are called generators) such that*

$$E \subseteq \bigcup_{i=1}^{m} (\underline{\alpha_i} + \mathbb{N}^n).$$

Observe that every monoideal has a set of generators (for example the whole monoideal).

Proposition 2.1.11. *Every monoideal E of $\mathbb{N}^n$ possesses a unique finite minimal set of generators B.*

Definition 2.1.12. An ***admissible ordering*** on $(\mathbb{N}^n, +)$ is a total monoid ordering $\preccurlyeq$ such that $\underline{0} \preccurlyeq \underline{\alpha}$ for every $\underline{\alpha} \in \mathbb{N}^n$. By remark 2.1.6 the total ordering $\preccurlyeq$ is admissible iff it satisfies the following two conditions:

(1) $\underline{0} \prec \underline{\alpha}$ for every $\underline{0} \neq \underline{\alpha} \in \mathbb{N}^n$;
(2) $\underline{\alpha} + \underline{\gamma} \prec \underline{\beta} + \underline{\gamma}$ for all $\underline{\alpha}, \underline{\beta}, \underline{\gamma} \in \mathbb{N}^n$ with $\underline{\alpha} \prec \underline{\beta}$.

The ***total degree*** of the element $\underline{\alpha} \in \mathbb{N}^n$ is

$$|\underline{\alpha}| = \alpha_1 + \cdots + \alpha_n.$$

Example 2.1.13. The ***total degree ordering*** $\preccurlyeq_{tot}$ on $\mathbb{N}^n$ is defined by

$$\underline{\beta} \preccurlyeq_{tot} \underline{\alpha} \Leftrightarrow (|\underline{\beta}| < |\underline{\alpha}|) \text{ or } (\underline{\beta} = \underline{\alpha}).$$

The ordering $\preccurlyeq_{tot}$ is only a partial ordering, and hence not an admissible ordering.

For any $1 \leq i \leq n$ we denote by ϵ_i the element $(0, \ldots, 1, \ldots, 0) \in \mathbb{N}^n$ whose all entries are 0 except for the value 1 in the i-th component.

Example 2.1.14. The ***reverse lexicographical ordering*** $\preccurlyeq_{revlex}$ on $\mathbb{N}^n$ with $\epsilon_1 \prec \epsilon_2 \prec \cdots \prec \epsilon_n$ is defined by

$$\underline{\alpha} \prec_{revlex} \underline{\beta} \Leftrightarrow \exists j \in \{1, 2, \ldots, n\} \text{ such that } \alpha_i = \beta_i \, \forall i < j \text{ and } \alpha_j > \beta_j.$$

The ordering $\preccurlyeq_{revlex}$ is a total ordering which is compatible with the monoid structure, but is not admissible since $\underline{0}$ is the biggest element w.r.t. this ordering.

Let us give now some examples of standard admissible orderings.

Example 2.1.15. The ***lexicographical ordering*** $\preccurlyeq_{lex}$ on $\mathbb{N}^n$ with $\epsilon_1 \prec \epsilon_2 \prec \cdots \prec \epsilon_n$ is defined by

$$\underline{\alpha} \prec_{lex} \underline{\beta} \Leftrightarrow \exists j \in \{1, 2, \ldots, n\} \text{ such that } \alpha_i = \beta_i \, \forall i > j \text{ and } \alpha_j < \beta_j.$$

Example 2.1.16. The ***degree lexicographical ordering*** $\preccurlyeq_{deglex}$ on $\mathbb{N}^n$ with $\underline{\epsilon_1} \prec \underline{\epsilon_2} \prec \cdots \prec \underline{\epsilon_n}$ is defined by

$$\underline{\alpha} \prec_{deglex} \underline{\beta} \text{ iff } \Leftrightarrow |\underline{\alpha}| < |\underline{\beta}| \text{ or } \left(|\underline{\alpha}| = |\underline{\beta}| \text{ and } \underline{\alpha} \prec_{lex} \underline{\beta}\right).$$

Example 2.1.17. The ***degree reverse lexicographical ordering*** $\preccurlyeq_{degrevlex}$ on $\mathbb{N}^n$ with $\underline{\epsilon_1} \prec \underline{\epsilon_2} \prec \cdots \prec \underline{\epsilon_n}$ is defined by

$$\underline{\alpha} \prec_{degrevlex} \underline{\beta} \Leftrightarrow |\underline{\alpha}| < |\underline{\beta}| \text{ or } \left(|\underline{\alpha}| = |\underline{\beta}| \text{ and } \underline{\alpha} \prec_{revlex} \underline{\beta}\right).$$

Example 2.1.18. Let $\underline{\omega} = (\omega_1, \ldots, \omega_n) \in \mathbb{N}^n$. The ***weighted total degree*** with respect to $\underline{\omega}$ of the element $\underline{\alpha} \in \mathbb{N}^n$ is

$$|\underline{\alpha}|_{\underline{\omega}} = \langle \underline{\omega}, \underline{\alpha} \rangle = \sum_{i=1}^{n} \omega_i \alpha_i.$$

The ***$\underline{\omega}$-weighted degree lexicographical ordering*** $\preccurlyeq_{\underline{\omega}}$ on $\mathbb{N}^n$ with $\underline{\epsilon_1} \prec \underline{\epsilon_2} \prec \cdots \prec \underline{\epsilon_n}$ is defined by

$$\underline{\alpha} \prec_{\underline{\omega}} \underline{\beta} \text{ iff } \Leftrightarrow |\underline{\alpha}|_{\underline{\omega}} < |\underline{\beta}|_{\underline{\omega}} \text{ or } \left(|\underline{\alpha}|_{\underline{\omega}} = |\underline{\beta}|_{\underline{\omega}} \text{ and } \underline{\alpha} \prec_{lex} \underline{\beta}\right).$$

Proposition 2.1.19. *Any admissible ordering $\preccurlyeq$ on $\mathbb{N}^n$ is a refinement of the partial ordering $\preccurlyeq^n$ (defined in 2.1.9), that is, $\underline{\alpha} \preccurlyeq^n \underline{\beta}$ implies $\underline{\alpha} \preccurlyeq \underline{\beta}$.*

Proposition 2.1.20. *Any admissible ordering on $\mathbb{N}^n$ is a well-ordering, that is, every non-empty subset of $\mathbb{N}^n$ has a least element.*

Remark 2.1.21. Note that, the set of standard monomials $\mathrm{Mon}(x_1, \ldots, x_n)$ can be identified with (a subset of) $\mathbb{N}^n$ via the correspondence $\mathbf{x}^{\underline{\alpha}} \longleftrightarrow \underline{\alpha}$. The ordering on $\mathrm{Mon}(x_1, \ldots, x_n)$ induced by a monoid ordering $\prec$ on $\mathbb{N}^n$ will be called a monomial ordering.

We shall call the ordering on monomials, induced by an admissible ordering on exponents, a ***global monomial ordering***, or by abuse of notation, an ***admissible monomial ordering***.

More general, non-admissible, local and mixed monomial orderings are considered in Chapter 4.

2.2 G-algebras

Definition 2.2.1. Let $T = \Bbbk\langle x_1, \ldots, x_n \rangle$ be the free associative $\Bbbk$-algebra. Let c_{ji} be nonzero elements from $\Bbbk$ and p_{ji} be standard polynomials from T, where $1 \leqslant i < j \leqslant n$. The $\Bbbk$-algebra generated by $x_1, \ldots, x_n$ over $\Bbbk$, subject to the following relations

$$\mathcal{F} := \{x_j \cdot x_i - c_{ji} x_i x_j - p_{ji} = 0\}_{1 \leqslant i < j \leqslant n},$$

together with an admissible monomial well-ordering $\prec$ on $\mathrm{Mon}(x_1, \ldots, x_n)$ is called a ***G-algebra***, if the following conditions hold:

2.2. G-ALGEBRAS

1. $\forall \mathbf{x}^{\underline{\alpha}} \in \text{Mon}(p_{ij}) : \mathbf{x}^{\underline{\alpha}} \prec x_i x_j$, for all $1 \leqslant i < j \leqslant n$,
2. For all $1 \leqslant i, j, k \leqslant n$, the following element is reduced to zero w.r.t. the relations $\mathcal{F}$ in T:

$$c_{ki}c_{kj} \cdot p_{ji} \cdot x_k - x_k \cdot p_{ji} + c_{kj} \cdot x_j \cdot p_{ki} - c_{ji} \cdot p_{ki} \cdot x_j + p_{kj} \cdot x_i - c_{ji}c_{ki} \cdot x_i \cdot p_{kj} \in T.$$

This G-algebra shall be denoted by

$$\Bbbk \langle x_1, \ldots, x_n \mid x_j \cdot x_i = c_{ji} x_i x_j - p_{ji}, \prec \rangle.$$

Definition 2.2.2. An algebra A shall be called a ***GR-algebra***, if there exists a surjective $\Bbbk$-algebra homomorphism from A onto a G-algebra, that is, if A is a quotient of a G-algebra by a two-sided (nonzero) ideal.

Remark 2.2.3. Algebras similar to the above, were first introduced (in a slightly different way) by J. Apel in [5] under the name of G-algebras and were further studied in [99]. After that they were reinvented, named differently and studied by a lot of people. These algebras are also called ***rings of solvable type*** and resp. ***algebras of solvable type*** in [76] and ***PBW-algebras*** in [19]. H. Kredel generalized (cf. [78]) the algebras of solvable type even further by allowing scalars to be non-commutative, although these algebras are not $\Bbbk$-algebras anymore.

In this thesis we adopt some notations and conventions from [86, 67].

Remark 2.2.4. Note that Condition 2 of Definition 2.2.1, which is equivalent to equalities $x_i \cdot (x_j \cdot x_k) = (x_i \cdot x_j) \cdot x_k$ for all $1 \leqslant i, j, k \leqslant n$ taking place in the G-algebra, assures the associativity of multiplication and the existence of a PBW basis in a G-algebra. This condition is due to [87, Section 2] and [86, Section 1.2], is called the ***non-degeneracy condition***, also corresponds to the overlap ambiguities of Bergman for being resolvable (cf. [14]) or the (Noetherian) rewriting system arising from $\mathcal{F}$ for being complete (cf. [77]). In particular, in the case of universal enveloping algebra of a Lie algebra, the non-degeneracy condition corresponds to the Jacobi identity for the Lie algebra.

Let, in what follows, A stand for a G-algebra in n variables $x_1, \ldots, x_n$, endowed with a (fixed) total ordering on its PBW-basis.

Proposition 2.2.5 (cf. Proposition 1.9.2 from [67]). *Let A be as above. Then*

1. *A is a PBW-algebra, with the following PBW-basis: $\text{Mon}(A) = \text{Mon}(x_1, \ldots, x_n)$, which is totally sorted w.r.t. $\preccurlyeq$. Hence the notions of leading monomial/exponent/-coefficient/term are applicable to G-algebras.*
2. *A is left and right Noetherian,*
3. *A is an integral domain.*

Definition 2.2.6. Let F be any subset of A.

- We denote by $\mathcal{L}(F)$ the monoid ideal in $(\mathbb{N}^n, +)$, generated by exponents of the leading monomials of elements of F:

$$\mathcal{L}(F) := \langle \mathrm{Exp}(f) \in \mathbb{N}^n \mid f \in F, f \neq 0 \rangle_{\mathbb{N}^n} \subset \mathbb{N}^n.$$

  The ideal $\mathcal{L}(F)$ is called the **monoid ideal of leading exponents**.
- The **set of leading monomials** of F, denoted by $\mathrm{L}(F)$, is the $\mathbb{k}$-vector space, spanned by monomials, divisible by leading monomial of some element from F, that is:

$$\mathrm{L}(F) := \langle \mathbf{x}^{\underline{\alpha}} \in \mathrm{Mon}(A) \mid \exists f \in F, f \neq 0 \text{ such that } \mathrm{Lm}(f) | \mathbf{x}^{\underline{\alpha}} \rangle_{\mathbb{k}} \subset A.$$

The ideal $\mathcal{L}(F)$ is finitely generated due to Dickson's Lemma (cf. 2.1.10). Note moreover that

$$\mathrm{L}(F) = \langle \mathbf{x}^{\underline{\alpha}} \in \mathrm{Mon}(A) \mid \underline{\alpha} \in \mathcal{L}(F) \rangle_{\mathbb{k}} \subset A.$$

Definition 2.2.7. Let $I \subset A$ be a left (resp. right, resp. two-sided) ideal in A and $G \subset I$ a finite subset. Then G is called a left (resp. right, resp. two-sided) **Gröbner basis** of I if for any $f \in I \setminus \{0\}$ there exists $g \in G \setminus \{0\}$ such that: $\mathrm{Lm}(g) | \mathrm{Lm}(f)$.

Proposition 2.2.8. *Let $I \subset A$ be a left ideal and $G \subset I$ a finite subset. Then the following conditions are equivalent:*

1. *G is a (left) Gröbner basis of I.*
2. *$\mathrm{L}(G) = \mathrm{L}(I)$.*
3. *$\mathcal{L}(G) = \mathcal{L}(I)$ as monoid ideals in $(\mathbb{N}^n, +)$.*

The notion of **division with remainder** in the non-commutative setting can be formalized via the following notion of (left) normal form.

Definition 2.2.9. Let $\mathcal{G}_A$ denote the set of all finite ordered subsets of A. A map

$$\mathrm{NF} : A \times \mathcal{G}_A \to A, (f, G) \mapsto \mathrm{NF}(f \mid G),$$

is called a **(left) normal form** on A if, for all $G \in \mathcal{G}_A, f \in A$,

1. $\mathrm{NF}(0 \mid G) = 0$,
2. $\mathrm{NF}(f \mid G) \neq 0 \Rightarrow \mathrm{Lm}(\mathrm{NF}(f \mid G)) \notin \mathrm{L}(G)$,
3. $f - \mathrm{NF}(f \mid G) \in {}_A\langle G \rangle$, and if $G = \{g_1, \ldots, g_s\}$ then $f - \mathrm{NF}(f \mid G)$ (or, by abuse of notation, f) has a **standard representation** with respect to G, that is,

$$f - \mathrm{NF}(f \mid G) = \sum_{i=1}^{s} a_i g_i, \; a_i \in A, s \geqslant 0, \qquad (2.3)$$

satisfying $\mathrm{Lm}(\sum_{i=1}^{s} a_i g_i) \geqslant \mathrm{Lm}(a_i g_i)$ for all i such that $a_i g_i \neq 0$.

2.2. G-ALGEBRAS

Proposition 2.2.10. *Let $I \subset A$ be a left ideal, G a left Gröbner basis of I and $\mathrm{NF}(\cdot \mid G)$ a left normal form on A with respect to G. Then*

1. *For any $f \in A$, we have: $f \in I \Leftrightarrow \mathrm{NF}(f \mid G) = 0$.*
2. *If $J \subset A$ is a left ideal with $I \subset J$, then $\mathrm{L}(I) = \mathrm{L}(J)$ implies $I = J$. In particular, G generates I as a left ideal.*

Definition 2.2.11. *Let $f, g \in A \setminus \{0\}$. The **left S-polynomial** of f and g is defined by*

$$\mathrm{LeftSPoly}(f, g) := a_1 v_1 \cdot f - a_2 v_2 \cdot g,$$

where $w = \mathrm{lcm}(\mathrm{Lm}(f), \mathrm{Lm}(g))$, $v_1 = w / \mathrm{Lm}(f)$, $v_2 = w / \mathrm{Lm}(g)$, $a_2 = \mathrm{Lc}(v_1 \cdot f)$, $a_1 = \mathrm{Lc}(v_2 \cdot g)$.

Remark 2.2.12. If $\mathrm{Lm}(f)$ divides $\mathrm{Lm}(g)$ then reductum of g by f is their S-polynomial $h := \mathrm{LeftSPoly}(f, g)$, with even smaller leading monomial: $\mathrm{Lm}(h) \prec \mathrm{Lm}(g)$ if $h \neq 0$. If we proceed further reducing g by suitable (dividing) $f \in F$ we can achieve a strictly descending sequence of leading monomials: $\mathrm{Lm}(g) \succ \mathrm{Lm}(h) \succ \ldots$ This procedure gives rise to the generic (left) normal form algorithm, which always terminates if $\prec$ is a well-ordering.

Theorem 2.2.13 (Left Buchberger's Criterion). *Let $I \subset A$ be a left ideal and $G = \{g_1, \ldots, g_k\}$, $g_i \in I$. Let $\mathrm{NF}(\cdot \mid G)$ be a left normal form on A with respect to G. Then the following statements are equivalent:*

1. *G is a Gröbner basis of I,*
2. *for all $f \in I$: $\mathrm{NF}(f \mid G) = 0$,*
3. *each $f \in I$ has a (left) standard representation w.r.t. G,*
4. *for all $1 \leqslant i, j \leqslant k$: $\mathrm{NF}(\mathrm{LeftSPoly}(g_i, g_j) \mid G) = 0$.*

We may summarize theorem 2.2.13, as follows:

Remark 2.2.14. A finite subset $G \subset A$ is a Gröbner basis iff all left S-polynomials are reducible to zero modulo G:

$$G \text{ is a GB} \Leftrightarrow \forall f, g \in G : \mathrm{NF}(\mathrm{LeftSPoly}(f, g) \mid G) = 0.$$

Remark 2.2.15. Theorem 2.2.13 gives rise to generic Buchberger's algorithm for computing a (left) GB of an ideal, which always terminates (since $\mathcal{L}(G)$ can only ascend up to $\mathcal{L}(I)$), provided the left normal form terminates.

Remark 2.2.16. Later in this thesis (Chapter 4) we will show that in some local cases one can compute, so called, *standard bases* (which are generalizations of GBs), simply by considering weak left normal forms in homogenized settings instead of left normal forms.

Recall that f is called **reduced w.r.t. F**, if no monomial of f is in $\mathrm{L}(F)$. Given any left normal form, it is easy to extend it to a reduced normal form algorithm (cf. Algorithm 2.2.1).

Algorithm 2.2.1 REDLEFTNF(f, F)

ASSUME: A is a GR algebra s.th. either $<$ is a well-ordering or both f and F are homogeneous, NF$(-\mid -)$ is any left normal form over A
INPUT: $f \in A, F \in \mathcal{G}_A$;
OUTPUT: $h \in A$, a reduced left normal form of f w.r.t. F.
1: $h := 0; g := f$;
2: **while** $g \neq 0$ **do**
3: $\quad g := \mathrm{NF}(g \mid F)$;
4: $\quad h := h + \mathrm{Lt}(g)$;
5: $\quad g := \mathrm{Tail}(g)$;
6: **end while**
RETURN: h;

Proposition 2.2.17. *Algorithm 2.2.1 terminates and computes a reduced left normal form of $f \in A$ with respect to $F \in \mathcal{G}_A$.*

Proof. The correctness of this algorithm follows from the definition of a reduced normal form, that is, if this algorithm terminates, the result is correct.

Assuming that the left normal form always terminates, Algorithm 2.2.1 terminates if either $<$ is a well-ordering or the input is homogeneous, since Tail(g) has strictly smaller leading monomial than g, for any non-zero g. ∎

Remark 2.2.18. Two-sided GB of an ideal in a GR-algebra A can be computed using by staring with the left ideal structure and *completing* (cf. [6]) it to the right ideal structure, while keeping the left one, which results in both left and right GB.

Let $F = \{f_1, \ldots, f_k\}$ be a minimal set of generators of a two-sided ideal I in A such that $_A\langle F\rangle = {_A\langle F\rangle}_A = I$. Then, due to [76], it follows that $\langle F\rangle_A = {_\mathcal{M}\langle F\rangle}_A = I$ and $_A\langle F\rangle = \langle F\rangle_A = I$.

We shall say that F is a ***two-sided GB*** of a two-sided ideal I, if it satisfies one of three conditions above.

Algorithm 3.1 from [86, Section 3.1] computes two-sided GBs.

2.3 Definitions of Gröbner Bases in quotient algebras

In Computer Algebra one usually starts by defining a reduction relation modulo a set of elements, which gives rise to a normal form (modulo the set). Due to the nature of reduction relations modulo a set of elements, one has a lot of freedom and often an element can have more than one normal form.

2.4. USUAL APPROACH TO COMPUTATIONS IN FACTOR ALGEBRAS

Luckily in the case of (polynomial) G-algebras definitions of a GB are equivalent to saying that reduction modulo a GB must solve an ideal membership problem, e.g. by Theorem 2.2.13:

$$\text{a finite set of elements } G \text{ is a GB} \iff \text{NF}(f \mid G) = 0, \forall f \in \langle G \rangle. \tag{2.4}$$

Moreover, for G-algebras the following implication

$$\text{NF}(f \mid F) = 0 \implies \text{NF}(g \cdot f \mid F) = 0, \forall g \in A \tag{2.5}$$

is a simple consequence of defining leading terms in such a way that $\text{Lm}(g \cdot f)$ is divisible by both $\text{Lm}(g)$ and $\text{Lm}(f)$.

Note that the implication (2.5) may fail in quotient algebras with zero-divisors, e.g. in an exterior algebra with variables $\xi_1 \ldots \xi_3$, endowed with a degree ordering $\text{Lm}(\xi_1 \cdot (\xi_1\xi_2 + \xi_3)) = \text{Lm}(\xi_1\xi_3)$ is not divisible by $\text{Lm}(\xi_1\xi_2 + \xi_3) = \xi_1\xi_2$. This has to be taken into account while defining a GB and characterizing it.

It was pointed out in [123] that there are at least two legitimate definition of a GB, one being equivalent to (2.4) and the other saying that reduction relation modulo a GB must be a confluent relation, which is equivalent to requiring a unique normal form modulo a GB. Confluent reduction relations are discussed in [14]. Clearly the first definition of a GB implies the second one. Moreover in all our application we actually need to solve ideal membership problems rather than have a unique normal form.

Note that we will use the definition due to [67] in terms of leading monomial ideals, which will be shown to be equivalent to the one given by (2.4), and later on (cf. Chapter 4) with a slight modification of reduction relation/normal form we will be able to compute in central localizations by using mixed monomial orderings, where it is known, that, even in the commutative case, no unique "normal form" exists.

2.4 Usual approach to computations in factor algebras

Let A be a G-algebra with a fixed ordering $\prec$, NF_A be a fixed normal form on A, I be a two-sided ideal in A with $P \subset A$ being its GB.

We consider the GR-algebra $R = A/I$, we set $\text{Mon}(R) := \text{Mon}(A) \setminus \text{L}(I)$ and assume that for each element $[f] = f + I \in R$ ($f \in A$) there is a representative, denoted by $\widetilde{f}$, which contains only terms with monomials being in $\text{Mon}(R)$.

Speaking about GBs in GR-algebras usually amounts to the following recipe (cf. [78, 76, 67, 86]), which steams from the desire to solve the ideal membership problem (cf. (2.4)) in these algebras:

Using the induced admissible monomial ordering on $\text{Mon}(R)$ we can define a reduction relation and a notion of GB via (2.4). Moreover for a finite subset $F \subset A$, such that,

$\forall f \in F : \mathrm{Lm}(f) \notin \mathrm{L}(I)$, and for any $g \in A$ it can be shown (under some mild assumption, cf. [72]) by induction on the number of reduction steps that

$$\mathrm{NF}_R([g] \mid [F]) = 0 \Longrightarrow \mathrm{NF}_A(g \mid F \cup P) = 0, \qquad (2.6)$$

and furthermore

$$F \cup P \text{ is a GB in } A \text{ iff } [F] \text{ is a GB in } R. \qquad (2.7)$$

Remark 2.4.1. Let A and R be as above, $[h]$ be an element in R, $\mathcal{J}$ be the left ideal in R generated by $G = \{[g_1], \ldots, [g_k]\}$. Denote by $\widetilde{\mathcal{J}}$ the left ideal in A generated by $\{\widetilde{g}_1, \ldots, \widetilde{g}_k\} \cup P$ and let $F = \{f_1, \ldots, f_l\}$ be its left Gröbner basis in A. Then the set $\{[\mathrm{NF}_A(f \mid P)] \mid f \in F\} \setminus \{[0]\}$ is a *left Gröbner basis* of $\mathcal{J}$ and the element $\mathrm{NF}_R([h] \mid G) := \left[\mathrm{NF}_A\left(\widetilde{h} \mid F\right)\right]$ is a *normal form* of $[h]$ w.r.t G.

Chapter 3

Gröbner bases in graded commutative algebras

Since graded commutative algebras can be represented as GR-algebras one can use the general GB algorithm for GR-algebras (see [5, 78, 19, 76, 86]) as explained in Section 2.4. This approach to graded commutative algebras has been used by D. Green in [62], whose treatment is explained in Section 3.1.

Let us remark that all commutative algebras can also be represented as factor algebras of free associative k-algebras but no-one uses this representation in practice by working directly with polynomials.

Similarly for graded commutative algebras we try to avoid the general quotient algebra approach (cf. Section 2.4). On the contrary, our approach is to make the most use of *a priori* knowledge about zero-divisors in graded commutative algebras.

In what follows we try to follow [67] as close as possible and highlight the differences occurring due to the presence of zero-divisors.

Our direct approach seems to be somewhat similar to that of [123, 102, 91]. But our characterization of a GB is by far much more direct, explicit and efficiently implementable.

We also give a short account of the results by T. Stockes (cf. [123]) and prove that his definition of a Gröbner Left Ideal Basis (GLIB) is equivalent to our definition of a GB of an ideal.

3.1 Green's approach

Throughout this section "graded" means $\mathbb{Z}$-graded.

As an example of the previously discussed approach (cf. 2.4) applied to graded commutative algebras let us interpret Chapter 4 from [62] (in the case of odd characteristic of k), using the standard computer algebra notions from [67, 86]. Note that the author's considerations for right ideals can be easily translated for left (one-sided) ideals by using the opposite algebra.

A *graded commutative algebra* $\mathcal{A}$ is an associative k-algebra and can be represented as the factor of a free associative (graded) k-algebra $\Bbbk\langle z_1,\ldots,z_n\rangle$ by the two-sided (graded) ideal in it, generated by relations $\{z_j \cdot z_i - (-1)^{t_i t_j} z_i \cdot z_j \mid 1 \leqslant i,j \leqslant n\}$, where $|z_i| := t_i \in \mathbb{Z}$ are degrees (grades) of algebra generators. These degrees also define the $\mathbb{Z}$-grading on $\mathcal{A}$.

From the defining relations it follows that $h^2 = 0$ for any odd homogeneous element of a graded commutative algebra. This holds even in the free graded commutative algebra.

By taking away zero-divisors from $\mathcal{A}$ we arrive to the following algebra:

Definition 3.1.1. Let $z_1,\ldots,z_n$ be variables, each equipped with a positive integer degree $|z_i| := t_i \in \mathbb{N}$.

1. The $\boldsymbol{\Theta}$-*algebra* $\mathcal{S} = \Theta(z_1,\ldots,z_n)$ on z_i over k is defined as the associative k-algebra
$$\Theta(z_1,\ldots,z_n) := \Bbbk\langle z_1,\ldots,z_n \mid z_j * z_i - (-1)^{t_i t_j} z_i * z_j = 0, 1 \leqslant i < j \leqslant n\rangle,$$
where the ordered n-tuple $(t_1,\ldots,t_n)$ is part of the structure of the Θ-algebra $\mathcal{S}$. Note that $\mathcal{S}$ is a polynomial algebra (that is, a G-algebra) with anti-commuting (of odd degree) and commuting (of even degree) variables but without zero-divisors. If we moreover denote its variables of even degree by $x_1,\ldots,x_n$ and variables of odd degree by $\xi_1,\ldots,\xi_m$ we may emphasize this this structure by writing $\Theta(x_1,\ldots,x_n;\xi_1,\ldots,\xi_m)$ Whenever exact grading is not important we may also write $\mathcal{S}^{n|m}$ for any $\Theta(x_1,\ldots,x_n;\xi_1,\ldots,\xi_m)$.

2. The family $\mathrm{Mon}(\mathcal{S})$ of monomials of $\mathcal{S}$ is defined as follows:
$$\mathrm{Mon}(\mathcal{S}) := \{\mathbf{z}^{\underline{\alpha}} := z_1^{\alpha_1} \cdots z_n^{\alpha_n} \mid (\alpha_1,\ldots,\alpha_n) \in \mathbb{N}^n\}.$$

3. for two monomials $\mathbf{z}^{\underline{\alpha}}, \mathbf{z}^{\underline{\beta}} \in \mathrm{Mon}(\mathcal{S})$ the **greatest common divisor** and the **least common multiple** are defined (respectively) as follows:
$$\gcd(\mathbf{z}^{\underline{\alpha}}, \mathbf{z}^{\underline{\beta}}) := \mathbf{z}^{\mathrm{Min}(\underline{\alpha},\underline{\beta})} \in \mathrm{Mon}(\mathcal{S}), \mathrm{lcm}(\mathbf{z}^{\underline{\alpha}}, \mathbf{z}^{\underline{\beta}}) := \mathbf{z}^{\mathrm{Max}(\underline{\alpha},\underline{\beta})} \in \mathrm{Mon}(\mathcal{S}),$$
where $\mathrm{Min}(\underline{\alpha},\underline{\beta})$ and $\mathrm{Max}(\underline{\alpha},\underline{\beta})$ are defined as component-wise Min and respectively Max.

Remark 3.1.2. Clearly any Θ-algebra is zero-divisors-free and the original graded commutative algebra can be represented as the following quotient algebra (i.e. as a GR-algebra):
$$\mathcal{A} = \mathcal{S}/\langle z_i \cdot z_i - (-1)^{t_i t_i} z_i \cdot z_i \mid 1 \leqslant i \leqslant n\rangle.$$

More specifically, the free graded commutative algebra with n even-degree variables $x_1,\ldots,x_n$ and m odd-degree variables $\xi_1,\ldots,\xi_m$ is the quotient of $\Theta(x_1,\ldots,x_n;\xi_1,\ldots,\xi_m)$ by the relations $\xi_j^2 = 0$:
$$\Theta(x_1,\ldots,x_n;\xi_1,\ldots,\xi_m)/\langle\xi_j^2 \mid 1 \leqslant j \leqslant m\rangle.$$

3.1. GREEN'S APPROACH

Whenever exact grading is not important we may also write $\mathcal{A}^{n|m}$ for any such graded commutative algebra.

Remark 3.1.3 (cf. Remark 4.1 from [62]). Due to David Green, computing a GB in a Θ-algebra instead of doing that in a graded commutative algebra has the following advantages:

- As far as possible, the "structural" relations $h^2 = 0$ are treated in exactly the same way as the remaining relations (i.e. the original ideal), which makes it easier to list the critical pairs.
- elements of a Θ-algebra may be represented as polynomials. The polynomial is unique and every polynomial can occur.
- GBs for one-sided ideals in a Θ-algebra are analogous to the usual commutative GBs, whereas GBs for two-sided ideals would be more complicated as they would share at least some of the characteristics of non-commutative GBs.

In what follows we examine the structure of Θ-algebras closely following [62, Chapter 4].

Remark 3.1.4 (remark 4.8 from [62]). Let us consider homogeneous (of degree 1) elements $y, y + z \in \Theta(y, z)$, then $(y + z) \cdot y = y^2 - y \cdot z$ is not equal to $y \cdot (y + z) = y^2 + y \cdot z$, even up to a sign.

In general, homogeneous elements of a Θ-algebra do not commute, even up to sign, in contrary to them doing so in the original graded commutative algebra (e.g. since $y^2 = 0$ holds there).

Since Θ-algebras do not have zero-divisors it makes sense to use any monomial orderings:

Definition 3.1.5 (Definition 4.9 from [62]). Let $\mathcal{S}$ be a Θ-algebra. A total ordering $\leq$ on $\text{Mon}(\mathcal{S})$ is called **usable** if from $\mathbf{z}^{\underline{\alpha}} \leq \mathbf{z}^{\underline{\beta}}$ follows that $\mathbf{z}^{\underline{\alpha}+\underline{\gamma}} \leq \mathbf{z}^{\underline{\beta}+\underline{\gamma}}$ for all $\underline{\alpha}, \underline{\beta}, \underline{\gamma} \in \mathbb{N}^n$.

Each element p of the Θ-algebra $\mathcal{S}$ has a finite support $\text{Supp}(p) \subseteq \text{Mon}(\mathcal{S})$ and as soon as we fix a *usable* (monomial) ordering on $\text{Mon}(\mathcal{S})$, each $p \neq 0$ has a leading monomial $\text{Lm}(p) \in \text{Mon}(\mathcal{S})$.

Definition 3.1.6. A *monomial ideal* in the Θ-algebra $\mathcal{S}$ is an ideal which is generated by a subset of the basis $\text{Mon}(\mathcal{S})$.

Note that for monomial ideals we do not need to distinguish between left, right and two-sided ideals since the left and the right ideal generated by a set of monomials coincide.

Proposition 3.1.7. *One-sided ideals in a Θ-algebra satisfy the ascending chain condition.*

Proposition 3.1.8. *Let $\mathcal{S} = \Theta(x_1, \ldots, x_n; \xi_1, \ldots, \xi_m)$ and $I \subset \mathcal{S}$ be a homogeneous one-sided ideal. If all $\xi_j^2 \in I$, then I is a two-sided ideal.*

Lemma 3.1.9. *Let* $\mathcal{S} = \Theta(x_1, \ldots, x_n; \xi_1, \ldots, \xi_m)$. *Let* $f_1, \ldots, f_r \in \mathcal{S}$ *be homogeneous of degree* d_f *and* $g_1, \ldots, g_r \in \mathcal{S}$ *of degree* d_g. *We denote* $\gamma := (-1)^{d_f d_g}$ *and*

$$\Phi := \sum_{i=1}^{r} (g_i \cdot f_i - \gamma f_i \cdot g_i).$$

Then there exist $h_1, \ldots, h_m \in \mathcal{S}$ *such that*

$$\Phi = \sum_{j=1}^{m} h_j \cdot \xi_j^2,$$

where each h_j *is either zero or homogeneous of degree* $d_f + d_g - 2$ *and* $\mathrm{Supp}(h_j \cdot \xi_j^2) \subseteq \mathrm{Supp}(\Phi)$ *for each* j.

The definition of a GB for one-sided ideals in Θ-algebras coincide with it for G-algebras.

D.Green gives the following definition for normal-form reductions in a Θ-algebra:

Proposition 3.1.10 (cf. [62], Proposition-Definition 4.27). *Let* $\mathcal{S}$ *be any* Θ-*algebra,* $I \subset \mathcal{S}$ *be a homogeneous one-sided ideal,* $f_S = (f_s \in I \mid s \in S)$ *a family of homogeneous non-zero elements and a usable (monomial) ordering* $\preccurlyeq$ *on* $\mathrm{Mon}(\mathcal{S})$. *Then the following holds:*

1. *For each* $f \in \mathcal{S}$ *there is an* $f' \in \mathcal{S}$, *an* $n > 0$ *and triples* $(s_i, m_i, \lambda_i)_{1 \leqslant i \leqslant n} \subset S \times \mathrm{Mon}(\mathcal{S}) \times \Bbbk^*$ *satisfying the following conditions:*

   (a) $f = \sum_{i=1}^{n} \lambda_i m_i f_{s_i} + f'$.

   (b) $\mathrm{Lm}(m_i f_{s_i}) \prec \mathrm{Lm}(m_j f_{s_j})$ *for* $1 \leqslant i < j \leqslant n$.

   (c) *if* $f' \neq 0$ *then* $\mathrm{Lm}(f') \notin \mathrm{L}(f_S)$ *and if moreover* $n > 0$ *then* $\mathrm{Lm}(f') < \mathrm{Lm}(m_n f_{s_n})$.

   Such an f' *shall be called a* **reduced form of** f **over** $(\preccurlyeq, f_S)$. *The following properties also hold:*

   (d) *if* $n > 0$ *then* f *is nonzero and* $\mathrm{Lt}(f) = \mathrm{Lt}(\lambda_1 m_1 f_{s_1})$

   (e) *if* $s_1, \ldots, s_r$ *are known for some* $r \leqslant n$ *then the* m_i *and the* λ_i *are uniquely determined for* $i \leqslant r$. *If* n *and* $s_1, \ldots, s_n$ *are known then* f' *is uniquely determined.*

2. *For each* $f \in \mathcal{S}$ *there is an* $f' \in \mathcal{S}$, *an* $n \geqslant 0$ *and triples* $(s_i, m_i, \lambda_i)_{1 \leqslant i \leqslant n} \subset S \times \mathrm{Mon}(\mathcal{S}) \times \Bbbk^*$ *satisfying the above conditions 1a, 1b and also* $\mathrm{Supp}(f') \cap \mathrm{L}(f_S) = \emptyset$. *Such an* f' *shall be called a* **completely reduced form of** f **over** $(\preccurlyeq, f_S)$. *Moreover Property 1e also holds here.*

Remark 3.1.11. Note that the reduced form (resp. completely reduced form) in proposition 3.1.10 is the usual normal form (resp. reduced normal form) from [67, 86], and it gives rise to the usual one-sided normal form (resp. reduced normal form) algorithm.

The only difference is that Proposition 3.1.10 does not require a global monomial ordering since it works in a homogeneous setting only: its argument for termination would be wrong

3.1. GREEN'S APPROACH

for non-homogeneous polynomials. Consider for instance, for the following (commutative) example: $A = \Bbbk[x], f = x, f_S = (x - x^2)$, w.r.t. the monomial ordering on $\text{Mon}(A)$ with $x \prec 1$. Clearly the sequence $f_0 = x, f_1 = x^2, \ldots$ is infinite whereas the sequence of its leading monomials is strictly decreasing w.r.t. the fixed local monomial ordering.

For any $f_s, f_t \in f_S$ there are unique terms τ_s, τ_t satisfying $\text{Lt}(\tau_s f_s) = \text{Lt}(\tau_t f_t)$ and $\text{Lm}(\tau_s f_s) = \text{lcm}(\text{Lm}(f_s), \text{Lm}(f_t)) = \text{Lm}(\tau_t f_t)$.

Definition 3.1.12. The *(left) S-polynomial* of f_s and f_t is defined by

$$\sigma_{st} := \sigma(f_s, f_t) := \tau_s f_s - \tau_t f_t.$$

Note that any monomial from $\text{Supp}(\sigma_{st})$ is smaller than $\text{lcm}(\text{Lm}(f_s), \text{Lm}(f_t))$ w.r.t. the fixed monomial ordering on $\text{Mon}(\mathcal{S})$.

Definition 3.1.13. 1. A homogeneous element $f \in \mathcal{S}$ is called **weakly reducible** if either $f = 0$ or there are $s_1, \ldots, s_r \in S$ and homogeneous elements $g_1, \ldots, g_r \in \mathcal{S} \setminus \{0\}$ satisfying the following conditions:
   (a) $f = \sum_{i=1}^{r} g_i f_{s_i}$,
   (b) $\text{Lm}(g_i f_{s_i}) \preccurlyeq \text{Lm}(f)$ for every i.

   In particular, each element having 0 as a reduced form is weakly reducible.

2. For two elements $f_s, f_t \in f_S$, the S-polynomial σ_{st} is called **weakly resolvable** if there are $s_1, \ldots, s_r \in S$ and homogeneous elements $g_1, \ldots, g_r \in \mathcal{S} \setminus \{0\}$ satisfying the following conditions:
   (a) $\sigma_{st} = \sum_{i=1}^{r} g_i f_{s_i}$,
   (b) $\text{Lm}(g_i f_{s_i}) \prec \text{lcm}(\text{Lm}(f_s), \text{Lm}(f_t))$ for every i.

   Clearly if σ_{st} is weakly reducible then it is weakly resolvable.

Remark 3.1.14. Clearly, a polynomial is weakly reducible iff it has a standard representation w.r.t. f_S.

Theorem 3.1.15 (Green's Buchberger's Criterion for Θ-algebras). *In the previously fixed setting we assume moreover that f_S generates I as a one-sided ideal. Then the following statements are equivalent:*

1. *f_S is a (one-sided) GB for I.*

2. *every reduced form of σ_{st} is zero for all $s, t \in S$.*

3. *σ_{st} has zero as a reduced form for all $s, t \in S$.*

4. *σ_{st} is weakly resolvable for all $s, t \in S$.*

If moreover ξ^2 is weakly reducible over f_S for every odd-dimensional generator $\xi \in \mathcal{S}$, then each of the following statements is equivalent to the first four statements.

5. Statement 2 holds for all s, t with $\gcd(\mathrm{Lm}(f_s), \mathrm{Lm}(f_t)) \neq 1$.

6. Statement 3 holds for all s, t with $\gcd(\mathrm{Lm}(f_s), \mathrm{Lm}(f_t)) \neq 1$.

7. Statement 4 holds for all s, t with $\gcd(\mathrm{Lm}(f_s), \mathrm{Lm}(f_t)) \neq 1$.

Remark 3.1.16. The fist four statements of theorem 3.1.15 is the usual equivalence which is true for any G-algebra (cf. [67, 86]).

On the other hand the last three statements mean that the usual Product Criterion is applicable in this homogeneous setting whenever squares of all odd-dimensional variables have standard representations w.r.t. the input ideal.

Theorem 3.1.17 (Buchberger's algorithm)**.** *Let* $\mathcal{S} = \Theta(x_1, \ldots, x_n; \xi_1, \ldots, \xi_m)$, *where each x_i is even and each ξ_j is odd-dimensional. Let $\preccurlyeq$ be a usable ordering on* $\mathrm{Mon}(\mathcal{S})$ *and I a homogeneous one-sided ideal in $\mathcal{S}$. Let $f_S = (f_s)_{s \in S}$ be a finite family of homogeneous nonzero elements of $\mathcal{S}$ which generates I.*

Assume further that all ξ_j^2 are weakly reducible over f_S (and therefore contained in I).

The Algorithm would be to proceed obtaining new family f_U by doing one of the following two steps as long as possible.

1. *Choose $s \neq t$ from S and compute a reduced form h over f_S of the S-polynomial σ_{st}. Set $T := S$. If $h = 0$ then set $U := S$, otherwise set $U := S \cup \{u\}$ and set $f_u := h$.*
   Qualification: *Only use pairs which were not yet considered before. Moreover, one may consider only pairs which satisfy $\gcd(\mathrm{Lm}(f_s), \mathrm{Lm}(f_t)) \neq 1$*
2. *Choose $s \in S$, set $T := S \setminus \{s\}$ and compute a reduced form h of f_s over f_T. Set $U := T$ if $h = 0$, otherwise set $U := T \cup \{u\}$ and $f_u := h$.*
   Qualification: *$\mathrm{Lt}(f_s)$ must be in $\mathrm{L}(f_T)$, that is, there should exist a term τ and $f_t \in f_T$ such that $\mathrm{Lt}(f_s) = \mathrm{Lt}(\tau f_t)$.*

In this process we obtain a sequence $f_{S(0)}, f_{S(1)}, \ldots$ by setting $S(0) = S$ and $f_{S(n+1)}$ to be the f_U for $f_{S(n)}$. This sequence has finite length and the last $f_{S(n)}$ is a minimal GB for I.

Remark 3.1.18. Theorem 3.1.17 gives the usual one-sided reduced Buchberger's algorithm (cf. [78, 19, 86]) in this homogeneous non-commutative setting. Due to Step 1 we build S-polynomials of pairs (of relatively prime leading monomials) and reduce everything reducible due to Step 2.

Thought David Green do not mention it, the assumption on the odd-dimensional generators in the theorem are necessary only in order to make the Product Criterion applicable in the case of graded commutative algebras, since the GB computation in a graded commutative algebra makes sure that squares of odd-dimensional generators belong to the family f_S.

3.2 Preliminaries

Remark. Since we work not necessary in a (graded) homogeneous setting, from the computational point of view it is only important to distinguish between commutative odd variables (we shall usually denote them $x_1, \ldots, x_n$) and anti-commutative even variables (denoted by $\xi_1, \ldots, \xi_m$).

Recall that a Θ-algebra generated by n even variables $x_1, \ldots, x_n$ and m odd variables $\xi_1, \ldots, \xi_m$ is the following PBW-algebra:

$$\mathcal{S}^{n|m} = \Bbbk\langle x_1, \ldots, x_n, \xi_1, \ldots, \xi_m \mid x_i x_j - x_j x_i = x_i \xi_k - \xi_k x_i = \xi_q \xi_p + \xi_p \xi_q = 0, p \neq q \rangle.$$

A standard monomial in n commutative variables $x_1, \ldots, x_n$ and m anti-commutative variables $\xi_1, \ldots, \xi_m$ is a power product

$$x^\alpha \xi^\beta = x_1^{\alpha_1} \cdot \ldots \cdot x_n^{\alpha_n} \xi_1^{\beta_1} \cdot \ldots \cdot \xi_m^{\beta_m} \in \mathcal{S}^{n|m},$$

where $\alpha = (\alpha_1, \ldots, \alpha_n) \in \mathbb{N}^n, \beta = (\beta_1, \ldots, \beta_m) \in \mathbb{N}^m$.

The set of standard monomials of $\mathcal{S}^{n|m}$ is denoted by

$$\text{Mon}(\mathcal{S}^{n|m}) := \text{Mon}(x_1, \ldots, x_n, \xi_1, \ldots, \xi_m) := \left\{ x^\alpha \xi^\beta \mid (\alpha, \beta) \in \mathbb{N}^n \times \mathbb{N}^m \right\}.$$

Remark 3.2.1. As usual for G-algebras one can identify $\text{Mon}(\mathcal{S}^{n|m})$ with $\mathbb{N}^n \times \mathbb{N}^m$, and thus a monomial ordering on $\text{Mon}(\mathcal{S}^{n|m})$ is induced by a total monoid order on $\mathbb{N}^n \times \mathbb{N}^m$.

By endowing $\mathcal{S}^{n|m}$ with a monomial ordering, $\mathcal{S}^{n|m}$ can be regarded as a G-algebra. That is, for any non-zero standard polynomial $p \in \mathcal{S}^{n|m}$ one can define all leading data functions (such as $\text{Lm}(p)$, $\text{Lt}(p)$, $\text{Lc}(p)$) in addition to the functions defined for any PBW-algebra: set of monomials occurring in p: $\text{Mon}(p)$ and their corresponding coefficients: $\text{Coef}(p, m)$.

Moreover, one already can compute GBs in $\mathcal{S}^{n|m}$ due to Section 2.2.

We denote the set of squares of odd-degree variables by $Q := \{\xi_1^2, \ldots, \xi_m^2\} \subset \mathcal{S}^{n|m}$. Clearly it is a two-sided GB of $\langle Q \rangle$ in $\mathcal{S}^{n|m}$. Therefore, the graded commutative algebra $\mathcal{A}^{n|m} = \mathcal{S}^{n|m}/\langle Q \rangle$ generated by n even variables $x_1, \ldots, x_n$ and m odd variables $\xi_1, \ldots, \xi_m$, can be regarded as a GR-algebra.

Proposition 3.2.2 (e.g. Theorem 3.2, [123])**.** *Graded commutative algebras are left, right and two-sided Noetherian. For instance, every left ideal is finitely generated.*

Observe that $\mathcal{A}^{n|m}$ is isomorphic to the tensor product of the commutative $\Bbbk$-algebra in n generators and the exterior algebra with m generators over $\Bbbk$:

$$\mathcal{A}^{n|m} \cong \Bbbk[x_1, \ldots, x_n] \otimes_{\Bbbk} \wedge(\xi_1, \ldots, \xi_m).$$

Remark 3.2.3. In order to turn the tensor product $A \otimes_{\Bbbk} B$ of two graded commutative algebras A and B into a graded commutative algebra with the same degrees of variables (i.e. $|a \otimes b| := |a| + |b|$) we need the following sign-twisted definition of multiplication:

$$(a \otimes b) \cdot (c \otimes d) := (-1)^{|b||c|+|a||d|} ac \otimes bd, \text{ where } a, c \in A, b, d \in B.$$

We denote this graded tensor product by $A \circledast_{\Bbbk} B$. It satisfies the proper graded commutativity relations:

$$(a \otimes b) \cdot (c \otimes d) = (-1)^{|a \otimes b||c \otimes d|} (c \otimes d) \cdot (a \otimes b), \text{ where } a, c \in A, b, d \in B.$$

In particular

$$(a \otimes 1) \cdot (1 \otimes b) = (-1)^{|a||b|} (1 \otimes b) \cdot (a \otimes 1),$$

which shows that both A and B are contained in it as subalgebras.

Similarly the $\mathbb{Z}_2$-graded tensor product of two super-commutative algebras is again a super-commutative algebra.

3.3 Direct approach

In this section we are going to develop the leading-data-notions directly for graded commutative algebras.

Let us consider a graded commutative algebra $\mathcal{A} := \mathcal{A}^{n|m}$. As noted before, it may be dealt with as a GR-algebra, that is, as the quotient algebra $\mathcal{S}/\langle Q \rangle$, where $\mathcal{S} := \mathcal{S}^{n|m}$ is an anti-commutative G-algebra or more precisely a Θ-algebra endowed with a monomial ordering. Let $Q := \{\xi_1^2, \ldots, \xi_m^2\} \subset \mathcal{S}$.

Let us denote the canonical projection of $f \in \mathcal{S}$ in $\mathcal{A}$ (considered as a quotient of $\mathcal{S}$) by $[f]$[1]. Since this projection is a homomorphism, the products in $\mathcal{S}$ and $\mathcal{A}$ are related by $[f] \cdot_{\mathcal{A}} [g] = [p \cdot_{\mathcal{S}} q]$, for any $p \in [f], q \in [g]$.

The canonical representative for the class $[f] \in \mathcal{A}$ can be obtained by killing monomials containing squares of anti-commutative variables in any polynomial g from $[f]$ (which is the same as computing the reduced normal form of g modulo Q).

Let $\mathcal{M} \subset \mathcal{S}$ be the $\Bbbk$-vector space spanned by

$$\left\{ x^\alpha \xi^\beta \mid \alpha \in \mathbb{N}^n, \beta \in \{0,1\}^m \right\} \subset \mathrm{Mon}(\mathcal{S}).$$

For every residue class $[q] \in \mathcal{A}$ there exists a unique canonical representative $p \in \mathcal{M} \cap [q]$. We denote it by $\widetilde{[q]}$ or simply by $\widetilde{q}$, that is, $\widetilde{q} := \widetilde{[q]} := p$.

Therefore, the map $\mathcal{A} \to \mathcal{M}, [q] \mapsto \widetilde{q}$ is an isomorphism of $\Bbbk$-vector spaces. Using this map we can translate the structure of $\Bbbk$-algebra from $\mathcal{A}$ to $\mathcal{M}$, that is, we introduce a new product rule on $\mathcal{M}$: $f \cdot_{\mathcal{M}} g := \widetilde{[f] \cdot_{\mathcal{A}} [g]} \in \mathcal{M}$, for $f, g \in \mathcal{M}$.

---
[1] Note that here and in what follows, we identify $[f] \in \mathcal{A}$ with the set $f + \langle Q \rangle \subset \mathcal{S}$

3.3. DIRECT APPROACH

We observe that
$$(\mathcal{M}, +_{\mathcal{S}}, \cdot_{\mathcal{M}}) \tag{3.1}$$
is a non-commutative $\Bbbk$-algebra isomorphic to $\mathcal{A}$.

Definition 3.3.1. Let
$$\mathcal{M} := {}_{\Bbbk}\langle x^{\alpha}\xi^{\beta} \mid \alpha \in \mathbb{N}^n, \beta \in \{0,1\}^m \rangle \subset \mathcal{S}.$$
We shall call the $\Bbbk$-algebra $(\mathcal{M}, +_{\mathcal{S}}, \cdot_{\mathcal{M}})$ (constructed above) the *graded commutative algebra representing $\mathcal{A}$*.

Now we can identify elements from $\mathcal{A}$ with their (polynomial) representatives in $\mathcal{M}$. Instead of an ideal J in $\mathcal{A}$ we shall now talk about an ideal in $\mathcal{M}$ given by canonical representatives of elements from J: $\left\{ \widetilde{f} \mid [f] \in \mathcal{J} \right\} =: \widetilde{J} \subset \mathcal{M}$, which is an ideal in $\mathcal{M}$ with respect to the newly introduced product on $\mathcal{M}$.

We choose as *PBW*-basis of $\mathcal{M}$ the set of standard monomials of $\mathcal{S}$ without squares of anti-commutative variables:
$$\mathrm{Mon}(\mathcal{M}) := \left\{ x^{\alpha}\xi^{\beta} \mid (\alpha, \beta) \in \mathbb{N}^n \times \{0,1\}^m \right\} \subset \mathrm{Mon}(\mathcal{S}). \tag{3.2}$$

Note that we will denote monomials from $\mathrm{Mon}(\mathcal{M})$ by z^{γ}, with $\gamma \in \mathbb{N}^n \times \{0,1\}^m$, instead of $x^{\alpha}\xi^{\beta}$, whenever this subdivision to odd/even variables is irrelevant.

Let us consider the product of two standard monomials $x^{\alpha}\xi^{(\beta_1,\ldots,\beta_m)}, x^{\gamma}\xi^{(\delta_1,\ldots,\delta_m)} \in \mathcal{M}$. There are two cases:

1. if there exists an index $1 \leq j \leq m$ such that $\beta_j = \delta_j = 1$ then $x^{\alpha}\xi^{\beta} \cdot_{\mathcal{M}} x^{\gamma}\xi^{\delta} = 0$,
2. otherwise
$$x^{\alpha}\xi^{\beta} \cdot_{\mathcal{M}} x^{\gamma}\xi^{\delta} = (-1)^s x^{\alpha+\gamma}\xi^{\beta+\delta}, \tag{3.3}$$
where $s = \sum_{j=1}^{m}\sum_{i=j+1}^{m} \beta_j \delta_i$, and the sums $\alpha+\gamma$ and $\beta+\delta$ are taken component-wise respectively in $\mathbb{N}^n$ and $\mathbb{N}^m$.

Remark 3.3.2. As in the case of G-algebra $\mathcal{S}$ (cf. 3.2.1) we may identify $\mathrm{Mon}(\mathcal{M})$ with $\mathbb{N}^n \times \{0,1\}^m \subset \mathbb{N}^n \times \mathbb{N}^m$. Unfortunately, in contrast to the usual case, the set of exponents $\mathbb{N}^n \times \{0,1\}^m$ is not a monoid with respect to component-wise addition in $\mathbb{N}^n \times \mathbb{N}^m$.

This can illustrated by observing that product of some monomials may be zero, which is not a monomial corresponding to the sum of exponents (e.g. $\xi \cdot_{\mathcal{M}} \xi = 0_{\mathcal{M}}$ cannot correspond to $(\mathbf{0}; 1) +_{\mathbb{N}^n \times \mathbb{N}^1} (\mathbf{0}; 1) = (\mathbf{0}; 2) \notin \mathbb{N}^n \times \{0,1\}$).

Moreover, if $\mathcal{M}$ contains zero-divisors (i.e. if $m > 0$) then there cannot exist a monomial ordering on $\mathrm{Mon}(\mathcal{M})$. Consider for example two anti-commutative variables ξ_1, ξ_2, and let $\xi_1 < \xi_2$ then $0 = \xi_1 \xi_1 < \xi_1 \xi_2 < \xi_2 \xi_2 = 0$, that is, $0 < 0$, which is a contradiction if we assume that a monomial ordering exists.

Definition 3.3.3. Let $<$ be a total-order on $\mathbb{N}^n \times \{0,1\}^m$ such that $\alpha + \delta < \beta + \delta$ whenever $\alpha < \beta$, provided $\alpha + \delta, \beta + \delta \in \mathbb{N}^n \times \{0,1\}^m$, where $\alpha, \beta, \delta \in \mathbb{N}^n \times \{0,1\}^m$ and sums are taken component-wise in $\mathbb{N}^n \times \mathbb{N}^m$ (which contains $\mathbb{N}^n \times \{0,1\}^m$).

Such a total ordering on $\mathbb{N}^n \times \{0,1\}^m$ induces a total order on $\mathrm{Mon}(\mathcal{M})$, which will be called a ***quasi-monomial ordering*** or simply *(monomial) ordering on $\mathcal{M}$*.

Any quasi-monomial ordering on $\mathrm{Mon}(\mathcal{M})$ satisfies the following condition: for any $z^\alpha, z^\beta, z^\delta \in \mathrm{Mon}(\mathcal{M})$ such that $z^\alpha \cdot_\mathcal{M} z^\delta$ and $z^\beta \cdot_\mathcal{M} z^\delta$ are non-zero (i.e. $\alpha + \delta, \beta + \delta \in \mathbb{N}^n \times \{0,1\}^m$), if $z^\alpha < z^\beta$ then $z^{\alpha+\delta} < z^{\beta+\delta}$.

Clearly, the monomial ordering on $\mathrm{Mon}(\mathcal{S})$ induces a quasi-monomial ordering on $\mathrm{Mon}(\mathcal{M})$. Furthermore, a global monomial ordering $>$ on $\mathrm{Mon}(\mathcal{S})$ is a well-ordering and thus the induced quasi-monomial ordering on $\mathrm{Mon}(\mathcal{M})$ is a well-ordering as well. By abuse of notation we will call such a quasi-monomial ordering on $\mathrm{Mon}(\mathcal{M})$ a ***global ordering***.

From now on we will always consider $\mathcal{M}$ to be equipped with an induced quasi-monomial ordering $>$ coming from $\mathrm{Mon}(\mathcal{S})$ as above. Since $\mathcal{M} \subset \mathcal{S}$ we can directly reuse the previously defined (for $\mathcal{S}$) notions of leading monomial/term/coefficient.

Definition 3.3.4. Let $z^\alpha, z^\beta \in \mathrm{Mon}(\mathcal{M})$. We say that z^α ***divides*** z^β, and denote this by $z^\alpha | z^\beta$, if $\alpha \leqslant_{nat} \beta$.

In such a case we know that the difference of exponents is again a valid exponent: i.e. $\beta - \alpha \in \mathbb{N}^n \times \{0,1\}^m$ and denote the corresponding quotient monomial by $z^\beta / z^\alpha := z^{\beta - \alpha} \in \mathrm{Mon}(A)$, where the difference of exponents is taken component-wise (say in $\mathbb{N}^n \times \mathbb{N}^m$).

We also denote $\mathrm{lcm}(z^\alpha, z^\beta) := z^{\mathrm{Max}(\alpha,\beta)} \in \mathrm{Mon}(\mathcal{M})$, where Max of two exponents is defined component-wise.

The next Remark 3.3.5 shows that a quasi-monomial ordering on $\mathrm{Mon}(\mathcal{M})$ is a well-order if and only if its restriction to the commutative monomials $x^\alpha \xi^0 \in \mathrm{Mon}(\mathcal{M})$ is a well-order:

Remark 3.3.5. If $>$ is a quasi-monomial ordering on $\mathrm{Mon}(\mathcal{M})$ then the following conditions are equivalent:

1. $>$ is a well-order.
2. $x_i > 1$ for $i = 1, \ldots, n$.
3. $x^\alpha > 1$ for all $\alpha \in \mathbb{N}^n \setminus \{(0,\ldots,0)\}$.
4. $\alpha \geqslant_{nat} \beta$ and $\alpha \neq \beta$ implies $x^\alpha \xi^\delta > x^\beta \xi^\delta$ for any $\delta \in \{0,1\}^m$, where $\alpha, \beta \in \mathbb{N}^n$.

Note that Conditions 2 and 3 are imposed only on commutative variables, resp. monomials, that is, we do not require any anti-commutative variables, resp. monomials to be greater then 1!

Proof. Implications $1 \Rightarrow 2 \Rightarrow 3 \Rightarrow 4$ can be shown as in the commutative case.

3.3. DIRECT APPROACH

For $4 \Rightarrow 1$ we do the following: let M be a set of monomials from $\mathcal{M}$ and for any $\gamma \in \{0,1\}^m$ let $M_\gamma := \{x^\alpha \mid x^\alpha \xi^\gamma \in M\}$. Then for any $\gamma \in \{0,1\}^m$ we are basically in the usual commutative case, that is, by Dickson's Lemma (cf. Lemma 2.1.10) there is a finite subset $B_\gamma \subset M_\gamma$ such that for any $x^\alpha \in M_\gamma$ there exists $x^\beta \in B_\gamma$ with $\beta \leq_{nat} \alpha$. By assumption: $x^\beta \xi^\gamma < x^\alpha \xi^\gamma$ or $x^\beta \xi^\gamma = x^\alpha \xi^\gamma$. Since there are only finitely many $\gamma \in \{0,1\}^m$, the element $\mathrm{Min}\bigl(x^\beta \xi^\gamma \mid x^\beta \in B_\gamma, \gamma \in \{0,1\}^m\bigr)$ is the smallest element of M, where the minimum is taken with respect to the ordering $>$ on $\mathrm{Mon}(\mathcal{M})$. ∎

Remark 3.3.6. If all $x_i > 1$ and $\xi_j > 1$ then (as in the commutative case) $>$ is a refinement of the natural partial ordering given by divisibility, that is, if $(\alpha, \gamma_1) \geq_{nat} (\beta, \gamma_2)$ and $(\alpha, \gamma_1) \neq (\beta, \gamma_2)$ then $x^\alpha \xi^{\gamma_1} > x^\beta \xi^{\gamma_2}$.

Proof. The conditions $(\alpha, \gamma_1) \geq_{nat} (\beta, \gamma_2)$ and $(\alpha, \gamma_1) \neq (\beta, \gamma_2)$ imply that $(\alpha - \beta, \gamma_1 - \gamma_2)$ is a non-zero valid exponent in $\mathbb{N}^n \times \mathbb{N}^m$ (or resp. in $\mathbb{N}^n \times \{0,1\}^m$). Thus $x^{\alpha-\beta} \xi^{\gamma_1 - \gamma_2} > 1$. Hence $x^\alpha \xi^{\gamma_1} = (\pm x^{\alpha-\beta} \xi^{\gamma_1 - \gamma_2}) x^\beta \xi^{\gamma_2} > x^\alpha \xi^{\gamma_2}$. ∎

Definition 3.3.7. Let F be any subset of $\mathcal{M}$. The ***leading ideal*** of F is the $\Bbbk$-vector space
$$\mathrm{L}(F) := {}_\Bbbk \langle m \in \mathrm{Mon}(\mathcal{M}) \mid \exists f \in F \text{ such that } \mathrm{Lm}(f) \text{ divides } m \rangle \subset \mathcal{M}.$$

Note that $\mathrm{L}(F)$ is a finitely generated monomial ideal in $\mathcal{M}$ (due to Dickson's Lemma).

Remark 3.3.8. Let F be any subset of $\mathcal{M}$. The monomial ideal $\mathrm{L}(F)$ has the following (almost tautological) properties:

1. $\forall 0 \neq f \in F$ it follows that $\mathrm{Lm}(f) \in \mathrm{L}(F)$,
2. if $F \subset G \subset \mathcal{M}$ then $\mathrm{L}(F) \subset \mathrm{L}(G)$, in particular, $\mathrm{L}(F) \subset \mathrm{L}(_\mathcal{M}\langle F \rangle)$.
3. $m \in \mathrm{L}(F)$ if and only if $\exists f \in F : \mathrm{Lm}(f) | \mathrm{Lm}(m)$.

In contrast to the case of G-algebras the equality

$$\mathrm{L}(\{f\}) = \mathrm{L}(_\mathcal{M}\langle f \rangle) \tag{3.4}$$

does not hold for all $f \in \mathcal{M}$ due to zero-divisors.

Consider, for example, $f := \xi_1 \xi_2 + 1 \in \mathcal{A}^{0|2} =: \mathcal{M}$. Then $g = \xi_1 \cdot f = \xi_1 \cdot (\xi_1 \xi_2 + 1) = \xi_1 \in {}_\mathcal{A}\langle f \rangle$ but $\mathrm{Lm}(g) = \xi_1$ is not divisible by $\mathrm{Lm}(f) = \xi_1 \xi_2$ and thus does not belong to $\mathrm{L}(\{f\})$. Proposition 3.3.12 below shows that the equality (3.4) will be true for special sets of generators, called *Gröbner bases*.

Definition 3.3.9. Let $I \subset \mathcal{M}$ be a left ideal in $\mathcal{M}$ and $G \subset I$ be a finite subset. Then G is called a ***left Gröbner basis*** (GB) of I if $_\mathcal{M}\langle G \rangle = I$ and for any $f \in I \setminus \{0\}$ there exists $g \in G$ satisfying $\mathrm{Lm}(g) | \mathrm{Lm}(f)$.

The right Gröbner basis is defined analogously. Two-sided GB is defined as before for G-algebras (cf. Remark 2.2.18).

Note that the "left Gröbner basis" constructed in Proposition 2.4.1 is a GB due to Definition 3.3.9.

Let us recall the reduction relation from [123]:

Definition 3.3.10. Let $F \subset \mathcal{M}$ be a finite subset, and $f, f' \in \mathcal{M}$ any elements. We say that f **left reduces** $f \to_F f'$ to f' modulo F (denoted by $f \to_F f'$) if $\exists p \in F$ and $\exists t \in \text{Mon}(f)$ such that $\text{Lm}(p)|t$ and $f' = f - (a/b)s \cdot p$, where $a = \text{Coef}(f,t), b = \text{Lc}(s \cdot p)$, $s = t/\text{Lm}(p) \in \text{Mon}(\mathcal{M})$ and in particular $\text{Lm}(s) \cdot \text{Lm}(p) \neq 0$.

Let $\to_F^*$ be the reflexive transitive closure of $\to_F$. Furthermore, the notions of reducibility, irreducibility and normal form modulo F can be introduced as usual. If an ordering on $\mathcal{M}$ is a well-order then a straightforward check shows that $\to_F^*$ is a Noetherian relation.

Remark 3.3.11. Let us figure out: what does $f \to_G^* 0$ mean? This means that there are $f_i \in {}_\mathcal{M}\langle \{f\} \cup G\rangle, g_i \in G$ and $a_i \in \mathcal{M}$, $i = 1, \ldots, k$ such that $f_i \to_{g_i} f_{i+1}$ (that is, $f_{i+1} = f_i - a_i \cdot g_i$), where $f_1 = f, f_{k+1} = 0$ and $\text{Lm}(a_i \cdot g_i) \in \text{Mon}(f_i)$ (in particular $\text{Lm}(f) = \text{Lm}(f_1) \geqslant \text{Lm}(f_2) \geqslant \cdots \geqslant \text{Lm}(f_k)$, $\text{Lm}(a_i \cdot g_i) \leqslant \text{Lm}(f_i)$ and $\text{Lm}(a_i) \cdot \text{Lm}(g_i) \neq 0$). By collecting all these reductions and putting them together we will get that f has the following representation with respect to G: $f = \sum_{i=1}^k a_i \cdot g_i$ and thus $f \in {}_\mathcal{M}\langle G\rangle$. It may be proven by induction on the number of reductions k that $\exists g \in G : \text{Lm}(g)|\text{Lm}(f)$.

Proposition 3.3.12. *Let $I \subset \mathcal{M}$ be a left ideal and $G \subset I$ a finite subset. Then the following conditions are equivalent:*

1. *G is a (left) Gröbner basis of I.*
2. *$\text{L}(G) = \text{L}(I)$.*
3. *For all $f \in \mathcal{M}$ the following holds true: $(f \in I \Leftrightarrow f \to_G^* 0)$.*

Proof. Condition 1 is equivalent to 2 as in the commutative case, since $\text{L}(-)$ has all needed properties due to remark 3.3.8:

(2) $\Leftarrow$ (1): Clearly, $\text{L}(G) \subset \text{L}(I)$ since $G \subset I$. Let $m \in \text{L}(I) \setminus \text{L}(G)$ then $\exists f \in I : \text{Lm}(f) = m$ and $\forall g \in G : \text{Lm}(g) \nmid m = \text{Lm}(f)$, which is a contradiction with G being a GB of I. Thus $\text{L}(I) = \text{L}(G)$.

(2) $\Rightarrow$ (1): Let $f \in I \setminus \{0\}$. Since $\text{L}(I) = \text{L}(G)$ and $\text{Lm}(f) \in \text{L}(I)$ it follows that $\text{Lm}(f) \in \text{L}(G)$. Hence there exists $g \in G : \text{Lm}(g)|\text{Lm}(f)$. Thus G is a GB of I.

(3) $\Leftarrow$ (1): Let G be a GB of I, we need to show both directions in 3.

$\Leftarrow$: If $f \to_G^* 0$ then from remark 3.3.11 it follows that $f \in {}_\mathcal{M}\langle G\rangle \subset I$.

$\Rightarrow$: Let us show that any $f \in I$ (left) reduces to zero modulo G: let $f_1 := f$, due to 1 there exists $g_1 \in G : \text{Lm}(g_1)|\text{Lm}(f_1)$ then we can reduce f_1 modulo g_1 to some $f_2 \in I$: $f_1 \to_{g_1} f_2$, and if $f_2 \neq 0$ we can go on. Thus we get the descending sequence of monomials $\text{Lm}(f) = \text{Lm}(f_1) \geqslant \text{Lm}(f_2) \geqslant \cdots \geqslant \text{Lm}(f_i) \geqslant \cdots$ which stabilizes at some i since $>$ is a well-ordering. If $f_i \neq 0$ then $\text{Lm}(f_i) \notin \text{L}(G) = \text{L}(I)$ which is a contradiction with $f_i \in I$. Thus f reduces to zero.

3.3. DIRECT APPROACH

(3) ⇒ (1): if $f \in I \setminus \{0\}$ then (by 3) $f \to_G^* 0$ and by remark 3.3.11 it follows that there exists $g \in G : \mathrm{Lm}(g) | \mathrm{Lm}(f)$. Thus G is a GB of I.

∎

Remark 3.3.13. Note that due to Proposition 3.3.12, our definition of a GB in a graded commutative algebra is equivalent to the definition of a **Gröbner Left Ideal Basis (GLIB)** from [123] (which is defined by Condition 3 of the proposition).

Definition 3.3.14. Let $\mathcal{G}_\mathcal{M}$ denote the set of all finite ordered subsets of $\mathcal{M}$. A map

$$\mathrm{NF} : \mathcal{M} \times \mathcal{G}_\mathcal{M} \to \mathcal{M}, (f, G) \mapsto \mathrm{NF}(f \mid G),$$

is called a **left normal form**normal form on $\mathcal{M}$ if, for all $G \in \mathcal{G}_\mathcal{M}, f \in \mathcal{M}$,

1. $\mathrm{NF}(0 \mid G) = 0$,
2. $\mathrm{NF}(f \mid G) \neq 0 \Rightarrow \mathrm{Lm}(\mathrm{NF}(f \mid G)) \notin \mathrm{L}(G)$,
3. $f - \mathrm{NF}(f \mid G) \in {}_\mathcal{M}\langle G \rangle$, and if $G = \{g_1, \ldots, g_s\}$ then $f - \mathrm{NF}(f \mid G)$ (or, by abuse of notation, f) has a **standard representation** with respect to G, that is,

$$f - \mathrm{NF}(f \mid G) = \sum_{i=1}^{s} a_i g_i, \ a_i \in \mathcal{M}, \tag{3.5}$$

satisfying $\mathrm{Lm}(\sum_{i=1}^s a_i g_i) \geqslant \mathrm{Lm}(a_i g_i)$ and

$$\mathrm{Lm}(a_i) \cdot \mathrm{Lm}(g_i) \neq 0 \tag{3.6}$$

for all i such that both a_i and g_i are non-zero.

Remark 3.3.15. Note that the additional condition (3.6) for nonzero a_i and g_i implies that $a_i \cdot g_i \neq 0$ and

$$\mathrm{Lm}(a_i \cdot g_i) = \mathrm{Lm}(\mathrm{Lm}(a_i) \cdot \mathrm{Lm}(g_i))$$

and thus $\mathrm{Lm}(g_i)$ divides $\mathrm{Lm}(a_i * g_i)$. Thus this important addition ensures that $\mathrm{Lm}(f - \mathrm{NF}(f \mid G))$ is divisible by some $\mathrm{Lm}(g_i)$ (if not zero).

Proposition 3.3.16. *Let $I \subset \mathcal{M}$ be a left ideal, G a (left) Gröbner basis of I and $\mathrm{NF}(\cdot \mid G)$ a left normal form on $\mathcal{M}$ with respect to G. Then*

1. *For any $f \in \mathcal{M}$, we have: $f \in I \Leftrightarrow \mathrm{NF}(f \mid G) = 0 \Leftrightarrow f \to_G^* 0$.*
2. *If $J \subset \mathcal{M}$ is a left ideal with $I \subset J$, then $\mathrm{L}(I) = \mathrm{L}(J)$ implies $I = J$. In particular, G generates I as a left ideal.*

Proof. The last part of condition 1 is due to proposition 3.3.12 ((1) implies (3)). Everything else is analogously to the commutative case by proposition 3.3.12 and definition 3.3.14:

(1) If $p := \mathrm{NF}(f \mid G) \neq 0$ then $\mathrm{Lm}(p) \notin \mathrm{L}(G) = \mathrm{L}(I)$, but since $f \in I$ and $G \subset I$ it follows that $p \in I$ and thus $\mathrm{Lm}(p) \in \mathrm{L}(I)$ (contradiction). For the other direction: if $\mathrm{NF}(f \mid G) = 0$ then $f \in {}_{\mathcal{M}}\langle G \rangle \subset I$. Thus $f \in I$.

(2) Since $G \subset I \subset J$ and $\mathrm{L}(J) = \mathrm{L}(I) = \mathrm{L}(G)$ it follows that G is a GB of J. Thus for $h \in \mathcal{M}$ we have that $h \in I \Leftrightarrow \mathrm{NF}(h \mid G) = 0 \Leftrightarrow h \in J$, and therefore $I = J$.

■

Proposition 3.3.17. *Let $I \subset \mathcal{M}$ be a left ideal and $G = \{g_1, \ldots, g_s\} \subset I$. Let $\mathrm{NF}_{\mathcal{M}}(- \mid G)$ be a left normal form on $\mathcal{M}$ with respect to G. Then G is a GB of I if and only if $\mathrm{NF}_{\mathcal{M}}(f \mid G) = 0$ for all $f \in I$.*

Proof. One direction has been already proven in 3.3.16. Now let $f \in I \setminus \{0\}$ satisfy $\mathrm{NF}_{\mathcal{M}}(f \mid G) = 0$, then due to standard representation of f w.r.t. G and our addition to it there exist $g \in G$ such that $\mathrm{Lm}(g) | \mathrm{Lm}(f)$, and thus G is a GB.

■

Definition 3.3.18. *Let $A, B \in \mathrm{Mon}(\mathcal{M})$ such that B divides A. Let us define a **signed quotient** of A and B by*
$$(A /\!/ B) := \sigma C,$$
where $C = A/B \in \mathrm{Mon}(\mathcal{M}), \sigma = \mathrm{Lc}(C \cdot B) \in \{\pm 1\}$.

Remark 3.3.19. It is easy to see that $(A /\!/ B) \cdot B = A$, whereas $(A/B) \cdot B = \pm A$. Moreover, if $A \cdot B = \sigma C$, where $A, B, C \in \mathrm{Mon}(\mathcal{M}), \sigma \in \{\pm 1\}$ then
$$A = \sigma\left(C /\!/ B\right) \tag{3.7}$$

Definition 3.3.20. *Let $f, g \in \mathcal{M} \setminus \{0\}$. The **(left) S-polynomial** of f and g is*
$$\mathrm{LeftSPoly}(f, g) := a_1 v_1 \cdot f - a_2 v_2 \cdot g,$$
where $w = \mathrm{lcm}(\mathrm{Lm}(f), \mathrm{Lm}(g)) \in \mathrm{Mon}(\mathcal{M})$ and
$$v_1 = w / \mathrm{Lm}(f), \quad v_2 = w / \mathrm{Lm}(g),$$
$$a_2 = \mathrm{Lc}(v_1 \cdot f), \quad a_1 = \mathrm{Lc}(v_2 \cdot g).$$

Note that the products of monomials $\mathrm{Lm}(v_1) \cdot \mathrm{Lm}(f)$ and $\mathrm{Lm}(v_2) \cdot \mathrm{Lm}(g)$ are non-zero by construction.

With the use of signed quotients the (left) S-polynomial of $f, g \in \mathcal{M}$ can be expressed as follows:
$$\mathrm{LeftSPoly}(f, g) = \mathrm{Lc}(g)\left(w /\!/ \mathrm{Lm}(f)\right) \cdot f - \mathrm{Lc}(f)\left(w /\!/ \mathrm{Lm}(g)\right) \cdot g.$$

Remark 3.3.21. The S-polynomial defined by Definition 3.3.20 has all the usual properties:

3.3. DIRECT APPROACH

- by construction $\text{Lm}(\text{LeftSPoly}(f, g)) < w$, if non-zero.
- if $\text{Lm}(g)$ divides $\text{Lm}(f)$ then $\text{Lm}(\text{LeftSPoly}(f, g)) < \text{Lm}(f)$, if non-zero.

Therefore the following usual left normal form Algorithm (3.3.1) works for graded commutative algebras.

Algorithm 3.3.1 LEFTNF(f, F)

ASSUME: $>$ is global or both f and F are graded homogeneous.
INPUT: $f \in \mathcal{M}, F \in \mathcal{G}_\mathcal{M}$;
OUTPUT: $h \in \mathcal{M}$, a left normal form of f with respect to F.
1: $h := f$;
2: **while** $((h \neq 0)$ and $(G_h := \{g \in F : \text{Lm}(g)|\,\text{Lm}(h)\} \neq \emptyset))$ **do**
3:    Choose any $g \in G_h$;
4:    $h := \text{LeftSPoly}(h, g)$;
5: **end while**
RETURN: h;

Proposition 3.3.22. *Algorithm 3.3.1 terminates and computes a left normal form of $f \in \mathcal{M}$ with respect to $F \in \mathcal{G}_\mathcal{M}$.*

Proof. Algorithm 3.3.1 terminates, as usual, since $\text{Lm}(h)$ drops on every iteration and this can not go on infinitely either in global case (*standard argument*: since the set of leading monomials must be well-ordered and thus has a minimum) or in graded homogeneous case, where degree of $h \neq 0$ would remain the same but there are only finitely many monomials of the same degree.

For the correctness of this algorithm we need to show that LeftNF has all needed properties of the definition 3.3.14: Property 1 is clear. Let $\text{LeftNF}_\mathcal{M}(f \mid F) =: h \neq 0$ then $G_h = \{g \in F : \text{Lm}(g)|\,\text{Lm}(h)\}$ is empty, and thus $\forall g \in F : \text{Lm}(g) \nmid \text{Lm}(h)$. It follows that $\text{Lm}(h) \notin \text{L}(F)$. For the last property 3 we only need to keep track of all reductions in S-polynomials as usual and notice that for any $g \in G_h$ we know that $\text{Lm}(g)$ divides $\text{Lm}(h)$ and thus the corresponding summand in standard representation (3.5) satisfy all needed conditions.

∎

Note that every specific choice of "any" $g \in G_h$ in Algorithm 3.3.1 (line 2) may lead to a different normal form.

3.4 Characterizations of Gröbner Bases

Remark 3.4.1. Let us recall from Remark 2.2.14 that a finite subset $F \subset \mathcal{S}$ is a Gröbner basis (in the Θ-algebra $\mathcal{S}$) if and only if all left S-polynomials reduce to zero modulo G:

$$G \text{ is a GB} \Leftrightarrow \forall f, g \in G : \mathrm{NF}(\mathrm{LeftSPoly}(f,g) \mid G) = 0. \tag{3.8}$$

In particular the set $Q = \{\xi_1^2, \ldots, \xi_m^2\} \subset \mathcal{S}$ is a GB since all S-polynomials are zero.

Remark 3.4.2. Unfortunately the above characterization of a GB (3.8) (in G-algebras) does not work in the case of a graded commutative algebra (and in particular, in an exterior algebra) because of zero-divisors.

Consider, for instance, the exterior algebra $\mathcal{A} = \mathcal{A}^{0|3}$ endowed with an orderings satisfying $\xi_1 > \xi_2 > \xi_3$ and let $g = \xi_1\xi_2 + \xi_3 \in \mathcal{A}$ then $F := \{g\} \subset \mathcal{A}^{0|3}$ is not a Gröbner basis since $f := \xi_1 \cdot g = \xi_1\xi_3 \in {}_\mathcal{A}\langle F \rangle$ but $\mathrm{Lm}(f) = \xi_1\xi_3$ is *not* divisible by $\mathrm{Lm}(g) = \xi_1\xi_2$ and thus cannot be reduced to 0 modulo g (the characterization (3.8) implies that any set containing a single element is a GB).

This example illustrates that because of zero-divisors in graded commutative algebras the following implication does not hold in contrary to a G-algebra case:

$$\forall g \in \mathcal{A}, f \in F \not\Rightarrow g \cdot f \to_F^* 0,$$

for any finite subset $F \subset \mathcal{A}$.

In the search for the best suited constructive characterization of a GB let us list some known ones, which may be used/reinterpreted in our setting.

Theorem 3.4.3 (Theorem 6.5 from [123]). *Let F be a finite subset of $\mathcal{M}$. The following statements are equivalent:*

1. *F is a GLIB (cf. Remark 3.3.13),*
2. *if $f_1, f_2 \in F$, then for any $t \in \mathrm{Mon}(\mathcal{M})$: $t \cdot f_1 \to_F^* 0$ and $t \cdot \mathrm{LeftSPoly}(f_1, f_2) \to_F^* 0$.*
3. *if $f_1, f_2 \in F$, $t_1 \in \mathrm{Mon}(\mathcal{M})$ satisfies $t_1 \cdot \mathrm{Lm}(f_1) = 0$ and $t_2 \in \mathrm{Mon}(\mathcal{M})$ satisfies $t_2 \cdot \mathrm{lcm}(\mathrm{Lm}(f_1), \mathrm{Lm}(f_2)) \neq 0$, then $t_1 \cdot f_1 \to_F^* 0$ and $t_2 \cdot \mathrm{LeftSPoly}(f_1, f_2) \to_F^* 0$.*

Remark 3.4.4. Condition 3 of Theorem 3.4.3 is equivalent to saying that for all $f, g \in F$, $m \in \mathrm{Mon}(\mathcal{M})$ the following holds

$$\begin{cases} \mathrm{NF}(m \cdot f \mid F) = 0, & \text{provided } m \cdot \mathrm{Lm}(f) = 0, \\ \mathrm{NF}(m \cdot \mathrm{LeftSPoly}(f,g) \mid F) = 0, & \text{provided } m \cdot \mathrm{lcm}(\mathrm{Lm}(f), \mathrm{Lm}(g)) \neq 0. \end{cases} \tag{3.9}$$

The following Proposition 3.4.5 illustrates that the quotient approach (cf. Section 2.3) is equivalent to ours.

3.4. CHARACTERIZATIONS OF GRÖBNER BASES

Proposition 3.4.5 (following Implication 15 and Theorem A.1 from [72]). *Let F be a finite subset from $\mathcal{M}$, $G = F \cup Q \subset \mathcal{S}$, then:*

1. $\forall g \in \mathcal{S}$ *it follows that: if* $\mathrm{NF}_{\mathcal{M}}(\widetilde{g} \mid F) = 0$ *then* $\mathrm{NF}_{\mathcal{S}}(g \mid G) = 0$,
2. F *is a GB in* $\mathcal{M}$ *if and only if* G *is a GB in* $\mathcal{S}$.

Proof. Statement 1 can be shown by induction on the number of reduction steps.

For Statement 2 we need to show both implications:

$\Leftarrow$ Let $f \in {}_{\mathcal{M}}\langle F \rangle \subset {}_{\mathcal{M}}\langle G \rangle \subset \mathcal{M} \subset \mathcal{S}$. Since G is a GB in $\mathcal{S}$ there exists $g \in G$: $\mathrm{Lm}(g) \mid \mathrm{Lm}(f)$, and since $\mathrm{Lm}(f) \in \mathcal{M}$ it follows that $\mathrm{Lm}(g) \in \mathcal{M}$ and thus $g \in \mathcal{M} \cap G = F$. Thus $g \in G \cap \mathcal{M} = F$ and F is a GB.

$\Rightarrow$ Let $g \in {}_{\mathcal{M}}\langle G \rangle \subset \mathcal{S}$ then $\widetilde{g} =: f \in {}_{\mathcal{M}}\langle F \rangle$, thus $\mathrm{NF}_{\mathcal{M}}(f \mid F) = 0$ since F is a GB in $\mathcal{M}$. Therefore $\mathrm{NF}_{\mathcal{S}}(g \mid G) = 0$ by 1. Hence G is a GB in $\mathcal{S}$ by Theorem 2.2.13.

∎

The following Theorem 3.4.6 is our characterization of a GB over graded commutative algebras, which seems to be the most direct and practically efficient approach. It is an easy corollary from our results about syzygies modules (Proposition 5.3.4 and Theorem 5.3.6) but can also be sketched by using the quotient approach (cf. [72, Theorem A.2]).

Theorem 3.4.6 (Characterization of a GB). *$F \in \mathcal{G}_{\mathcal{M}}$ is a GB in $\mathcal{M}$ if and only if the following conditions are satisfied:*

1. $\forall f \in F$ *and for all anti-commuting variables ξ_i dividing $\mathrm{Lm}(f)$ the following holds:*

$$\mathrm{NF}_{\mathcal{M}}(\xi_i \cdot f \mid F) = 0,$$

2. $\forall f, g \in F : \mathrm{NF}_{\mathcal{M}}(\mathrm{LeftSPoly}(f, g) \mid F) = 0$

Proof. Clearly, if F is a GB then the conditions 1 and 2 hold true.

Let us sketch the reverse implication by using the quotient approach (Proposition 3.4.5), due to which we need to show that $G = F \cup Q \in \mathcal{G}_{\mathcal{S}}$ is a GB in $\mathcal{S}$ by Buchberger's Criterion (Theorem 2.2.13). We consider 3 cases:

- If $f, g \in F$ then $\mathrm{NF}_{\mathcal{M}}(\mathrm{LeftSPoly}(f, g) \mid F) = 0$ due to Condition 2. Thus due to Proposition 3.4.5:

$$\mathrm{NF}_{\mathcal{S}}(\mathrm{LeftSPoly}(f, g) \mid G) = 0$$

- S-polynomials of pairs of elements from $Q \subset \mathcal{S}$ are all zeroes:

$$\mathrm{LeftSPoly}(\xi_i^2, \xi_j^2) = 0, \quad \forall \xi_i^2, \xi_j^2 \in Q.$$

- Only the following pairs remain: $(f, \xi_i^2) \in F \times Q$. Depending on whether $\mathrm{Lm}(f)$ contains a variable ξ_i or not we have one of the following cases:
  - If $\xi_i \nmid \mathrm{Lm}(f)$ then
  $$\mathrm{LeftSPoly}(f, \xi_i^2) = \mathrm{Tail}(\xi_i^2 \cdot_\mathcal{S} f)$$
  and it further reduces to 0 modulo $\{\xi_i^2\}$ in $\mathcal{S}$.
  - Otherwise, if $\xi_i \mid \mathrm{Lm}(f)$ then
  $$\mathrm{LeftSPoly}(f, \xi_i^2) = \mathrm{Tail}(\xi_i \cdot_\mathcal{S} f) := p.$$
  The polynomial $p \in \mathcal{S}$ reduces to $\xi_i \cdot_\mathcal{M} f$ modulo Q and due to Condition 1 is reducible to 0 modulo F. Thus $\mathrm{NF}_\mathcal{S}(\mathrm{LeftSPoly}(f, \xi_i^2) \mid G) = 0$ due to Proposition 3.4.5.

∎

Proposition 3.4.7. *Algorithm 3.4.1 terminates and computes a left GB of $F \in \mathcal{G}_\mathcal{M}$ with respect a global ordering on $\mathcal{M}$.*

Proof. Termination can be see as usual: since $\mathcal{M}$ is Noetherian the increasing sequence of leading monomial ideals $\mathrm{L}(G) \subset \mathcal{M}$ must stabilize at some step, which means that all needed leading monomials have been found, and thus Theorem 3.4.6 implies the correctness of the algorithm. ∎

Algorithm 3.4.1 LeftGB(F)

ASSUME: $\mathrm{NF}(- \mid -)$ is a normal form on $\mathcal{M}$
INPUT: $F \in \mathcal{G}_\mathcal{M}$.
OUTPUT: A left Gröbner basis G of $\mathrm{L}(F)$.
1: $L = \{(0, 0, f) \mid f \in F\}$; // initial pairset
2: $G = \emptyset$; // temporary GB
3: **while** $L \neq \emptyset$ **do**
4:     Choose and remove (p_1, p_2, h) from L;
5:     $h := \mathrm{NF}(h \mid G)$;
6:     **if** $h \neq 0$ **then**
7:         $L = L \cup \{(g, h, \mathrm{LeftSPoly}(g, h)) \mid g \in G\}$;
8:         $G = G \cup \{h\}$;
9:         **for all alternating variables** ξ dividing $\mathrm{Lm}(h)$ **do**
10:             $f = \xi *_M h$; // instead of $\mathrm{LeftSPoly}(\xi^2, h)$
11:             **if** $f \neq 0$ **then**
12:                 $L = L \cup \{(0, 0, f)\}$; // instead of (ξ^2, h, f)
13:             **end if**
14:         **end for**
15:     **end if**
16: **end while**
RETURN: G;

3.5 Criteria

The most time consuming part of the standard BBA is the reduction of S-polynomials. Therefore it is important to detect and throw away (without too much effort) those pairs whose S-polynomials *a priori* are reducible to zero and therefore useless for constructing a GB. Such Criteria are a part of every implementation of GB algorithm.

Bruno Buchberger gave two criteria (cf. [17, 18]) to detect pairs leading to unnecessary reductions in the commutative case:

- The **Product Criterion** says that LeftSPoly(f,g) can be reduced to 0 w.r.t. $\{f,g\}$, if $\gcd(\mathrm{Lm}(f), \mathrm{Lm}(g)) = 1$.
- The **Chain Criterion** says that given three pairs (f_i, f_j), (f_i, f_k) and (f_j, f_k) such that $\mathrm{Lm}(f_j)$ divides $\mathrm{lcm}(\mathrm{Lm}(f_i), \mathrm{Lm}(f_k))$, the pair (f_i, f_k) is superfluous as the corresponding S-polynomial LeftSPoly(f_i, f_k) is reducible to zero w.r.t. $\{f_i, f_j, f_k\}$. This criterion applies without restrictions to any module over any G- and GR-algebra (cf. [5, 86]).

Further criteria have been investigated by many people for the commutative case (cf. [46, 47, 56, 20, 96]). It would be interesting to generalize some of them to our noncommutative setting.

Remark 3.5.1. It was noted previously in Section 3.1 that the Product Criterion holds for graded homogeneous polynomials in Θ-algebras and thus can be used by Buchberger's algorithm in graded commutative algebras.

Let us give a simple counterexample showing that the product criterion does not hold for $\mathbb{Z}_2$-inhomogeneous elements in graded commutative algebras.

Example 3.5.2. Let $\mathcal{A} := \mathcal{A}^{2|2}$ and $f_i := x_i - \xi_i \in \mathcal{A}, i = 1,2$. Obviously f_i are $\mathbb{Z}_2$-inhomogeneous. Let $>$ be any ordering on $\mathcal{A}$ such that $x_1 > x_2 > \xi_1 > \xi_2$ (e.g. appropriate lexicographic ordering).

We claim that although $\gcd(\mathrm{Lm}(f_1), \mathrm{Lm}(f_2)) = 1$, the S-polynomial LeftSPoly(f_1, f_2) cannot be reduced to zero w.r.t. $\{f_1, f_2\}$.

Let us compute that S-polynomial: LeftSPoly$(x_1 - \xi_1, x_2 - \xi_2) = x_2(x_1 - \xi_1) - x_1(x_2 - \xi_2) = x_1\xi_2 - x_2\xi_1$. Its reduction can only go as follows: $x_1\xi_2 \to_{x_1-\xi_1} x_1\xi_2 - \xi_2(x_1 - \xi_1) = \xi_2\xi_1 = -\xi_1\xi_2$, $x_2\xi_1 \to_{x_2-\xi_2} x_2\xi_1 - \xi_1(x_2 - \xi_2) = \xi_1\xi_2$. Thus LeftSPoly$(f_1, f_2) \to^*_{\{f_1,f_2\}} (-\xi_1\xi_2) - (\xi_1\xi_2) = -2\xi_1\xi_2 =: p$. Since $\mathrm{Lm}(f_i) = x_i$ does not divide $\xi_1\xi_2 = \mathrm{Lm}(p)$ for any i it follows that p cannot be reduced any further. Hence LeftSPoly$(f_1, f_2) \not\to^*_{\{f_1,f_2\}} 0$. ∎

Remark 3.5.3. It was shown in [5] that the Chain Criterion holds for GR-algebras, and thus can be used by Buchberger's algorithm in graded commutative and super-commutative algebras.

We also give our generalized Chain Criterion in Lemma 5.3.7 which makes sense whenever we compute syzygies as well.

3.6 Kernel and preimage of a graded homomorphism

It is well known in the commutative case (e.g. cf. [67]) that the kernel and preimage of a map $A/I \to B/J$ between any quotient algebras can be computed by lifting the map to a map between polynomial algebras $A \to B$, considering the evaluation map $A \otimes_\Bbbk B \to B/J$ and computing its preimage by the elimination procedure. This approach can even be extended to some GR-algebras as shown in [86].

The case of homomorphisms between quotient algebras $\psi : A/I \to B/J$ is considered in [62], where A and B are two Θ-algebras and I, J are graded two-sided ideals. As it was noted there, it is not in general possible to lift ψ to a map of graded algebras $\Psi : A \to B$.

Example 3.6.1 (Example 4.34 from [62]). Let $A = B = \Theta(y, z)$ graded by $|y| := |z| := 1$. That is A and B are two anti-commutative polynomial G-algebras.

Let $I = J = \langle y^2, z^2 \rangle$ be graded two-sided ideals in A and B respectively.

Consider the map of graded algebras $\phi : A/I \to B/J$ given by $[y] \mapsto [y+z], [z] \mapsto [z]$. Recall that $[a]$ denotes the class $a + I$ for an element $a \in A/I$.

The only possible lift $\Phi : A \to B$ sends y to $y + z$ and z to z, but it does not respect the defining relation of A: $z \cdot y = -y \cdot z$ as Remark 3.1.4 has already shown.

Moreover the evaluation map $\pi : C \to B/J$ is generally not a map of graded algebras, where $C := A \otimes_\Bbbk B$ denotes the graded tensor product over $\Bbbk$, so that C is again a graded $\Bbbk$-algebra with "the same" grading, which seems to be implicitly defined by David Green for Θ-algebras as follows: $\Theta(y_i) \otimes_\Bbbk \Theta(z_j) := \Theta(y_i, z_j)$.

Example 3.6.2 (Example 4.35 from [62]). Let $A = \Theta(y), B = \Theta(z)$ graded by $|y| := 1$, let $I = J = 0$ and $\phi : y \mapsto z$. Then the graded tensor product $C := A \otimes_\Bbbk B = \Theta(y, z)$, with the same grading: $|y| := |z| := 1$, has the relation $yz = -zy$. But it can not be preserved under any map $C \to B$ sending both y and z to $z \in B/J$.

However, the aforementioned procedure works in the case when A/I and B/J are both graded commutative algebras (cf. [62, Section 4.3]).

Lemma 3.6.3 (Lemma 4.35 from [62]). *Let $A = \Theta(z_1, \ldots, z_n)$ and $B = \Theta(w_1, \ldots, w_m)$ be Θ-algebras, let $I \subset A, J \subset B$ be two-sided ideals and let $\phi : A/I \to B/J$ be a homomorphism of graded algebras. Let C be the Θ-algebra $\Theta(w_1, \ldots, w_m, z_1, \ldots, z_n)$ which contains both A and B as subalgebras.*

If B/J is graded commutative then the map

$$\nu : C \to B/J, \quad w_j \mapsto [w_j], \quad z_i \mapsto \phi([z_i])$$

is a map of graded algebras (i.e. relations are preserved), with the kernel

$$M := \langle b - a \mid a \in A, b \in B \text{ such that } \phi([a]) = [b] \rangle \subset C.$$

3.6. KERNEL AND PREIMAGE OF A GRADED HOMOMORPHISM

Moreover the kernel of the map ϕ between graded commutative algebras can be found by elimination of $b \in B$ from M:

$$A \cap M = \langle a \in A \mid [a] \in \mathrm{Ker}(\phi) \rangle.$$

The intersection $A \cap M$ can be computed via Buchberger's algorithm over the G-algebra C with respect to special eliminating ordering as shown in [62, Corollary 4.37].

Chapter 4

Localization

In order to be able to compute in central localizations of non-commutative algebras we allow commutative variables to be "local", generalize Mora algorithm (in a similar fashion as G.-M. Greuel and G. Pfister by allowing local or mixed monomial orderings, cf. [67]) and work with SBs instead of GBs.

The main theoretical result is Proposition 4.3.1. It allows us to extend non-commutative computer algebra to central localizations of GR-algebras by means of rings associated to monomial orderings. In order to show that our Mora normal form algorithm 4.4.1, works over these localizations, we prove that any GR-algebra can be appropriately homogenized (Proposition 4.4.7).

4.1 The commutative localization

In this section we recall the definition and some properties of localization of commutative rings and commutative rings associated to monomial orderings following [67, Sections 1.4 and 1.5].

- Let A be a commutative ring with a unit.
- Let for the set $S \subset A$ the following conditions hold:
  1. $1 \in S$,
  2. $a \in S$ and $b \in S$ implies that $a \cdot b \in S$.

  Such a set is called **multiplicative** or **multiplicatively closed**.
- Consider the equivalence relation on $A \times S$, defined by:

$$(a,b) \sim (a',b') \text{ iff } \exists s \in S \text{ such that } s(ab' - a'b) = 0.$$

- Denote by $\frac{a}{b}$ the equivalence class of $(a,b) \in A \times S$ w.r.t. $\sim$.

- The (commutative) **localization** or the **ring of fractions** of A w.r.t. S, denoted by S^1A, is defined as follows:
$$S^{-1}A := \left\{ \frac{a}{b} \,\Big|\, a \in A, b \in S \right\},$$
together with the ring operations defined by
$$\frac{a}{b} + \frac{a'}{b'} := \frac{ab' + a'b}{bb'}, \quad \frac{a}{b} \cdot \frac{a'}{b'} := \frac{aa'}{bb'}.$$

Example 4.1.1. Let P be a prime ideal in A. Clearly $A \setminus P$ is multiplicatively closed. We denote the localization of A w.r.t. $A \setminus P$ by A_P and call it the *localization* of A at the prime ideal P.

Let S be the set of all non-zero-divisors of A, which is clearly multiplicatively closed. For this S, we denote $S^{-1}A$ by $\mathrm{Quot}(A)$ and call it the **total ring of fractions** or the **total quotient ring** of A.

Proposition 4.1.2 (Proposition 1.4.5). *Let A, S and $S^{-1}A$ be as above, then:*

1. *The ring operations on $S^{-1}A$ are independent of the choice of representatives and thus well-defined.*
2. *The set $S^{-1}A$ together with the operations $\cdot$ and $+$ is a commutative ring with $1 = \frac{1}{1}$.*
3. *The (canonical) map $\jmath : A \to S^{-1}A : a \mapsto \frac{a}{1}$ is a ring homomorphisms, satisfying the following conditions:*
   - *(a) $\jmath(s)$ is a unit in $S^{-1}A$ if $s \in S$,*
   - *(b) $\jmath(a) = 0$ iff $as = 0$ for some $s \in S$,*
   - *(c) $\jmath$ is injective iff S has no zero-divisors,*
   - *(d) $\jmath$ is bijective iff S consists of units.*
4. *$S^{-1}A = 0$ iff $0 \in S$.*
5. *$S_1 \subset S_2$ are multiplicatively closed in A and have no zero-divisors then $S_1^{-1}A \subset S_2^{-1}A$.*
6. *Every ideal in $S^{-1}A$ is generated by the image of an ideal in A under the map $\jmath$. Moreover, the prime ideals in $S^{-1}A$ are in one-to-one correspondence with the prime ideals in A which do not meet S.*

Proposition 4.1.3 (Proposition 1.4.7). *Let A and B be two commutative unital rings, $\varphi : A \to B$ be a ring homomorphisms (which maps 1_A to 1_B), $S \subset A$ multiplicatively closed, and $\jmath : A \to S^{-1}A$ the canonical ring homomorphisms $a \mapsto \frac{a}{1}$.*

1. *If $\varphi(s)$ is a unit in B for all $s \in S$ then the following* universal *property holds: there exists a unique ring homomorphisms $\psi : S^{-1}A \to B$ which makes the following diagram commute:*

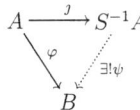

2. If moreover φ also satisfies the following conditions:
   (a) $\varphi(a) = 0$ implies $sa = 0$ for some $s \in S$,
   (b) every element of B is of the form $\varphi(a)\varphi(s)^{-1}$

   then φ is an isomorphism.

Let us now consider localizations of a commutative polynomial ring $A = \Bbbk[x_1,\ldots,x_n]$, endowed with monomial ordering $>$ (which is not necessary a well-ordering) w.r.t.

$$S_> := \{u \in A \setminus \{0\} \mid \mathrm{Lm}(u) = 1\},$$

which is multiplicatively closed due to the following properties of leading monomials: $\forall f,g \in A \setminus \{0\} : \mathrm{Lm}(fg) = \mathrm{Lm}(\mathrm{Lm}(f)\mathrm{Lm}(g)), \mathrm{Lm}(f+g) \leqslant \mathrm{Max}(\mathrm{Lm}(f),\mathrm{Lm}(f))$.

Denote **the ring associated to $\Bbbk[x]$ and $>$** by

$$\Bbbk[x]_> := S_>^{-1}\Bbbk[x].$$

Lemma 4.1.4 (Lemma 1.5.2). *Let $\Bbbk[x] = \Bbbk[x_1,\ldots,x_n]$, let $>$ be a monomial ordering on $\mathrm{Mon}(x_1,\ldots,x_n)$. Then*

1. $\Bbbk[x] \subset \Bbbk[x]_> \subset \Bbbk[x]_{\langle x \rangle}$.
2. *The set of units in $\Bbbk[x]_>$ is given by*

$$\left(\Bbbk[x]_>\right)^* = \left\{\frac{u}{v} \mid u,v \in \Bbbk[x], \mathrm{Lm}(u) = \mathrm{Lm}(v) = 1\right\},$$

and satisfies $\left(\Bbbk[x]_>\right)^* \cap \Bbbk[x] = S_>$.
3. $\Bbbk[x] = \Bbbk[x]_>$ *iff $>$ is a global ordering.*
4. $\Bbbk[x]_{\langle x \rangle} = \Bbbk[x]_>$ *iff $>$ is a local ordering.*
5. $\Bbbk[x]_>$ *is a Noetherian ring.*
6. $\Bbbk[x]_>$ *is factorial.*

Lemma 4.1.4 shows that by choosing a local ordering, we can, basically, do the calculations in the localization of a polynomial ring by choosing a mixed ordering ordering and compute with polynomials instead of rational functions. In particularly, we can efficiently compute in $\Bbbk[x_1,\ldots,x_n]_{\langle x_1,\ldots,x_k \rangle}$ for $k \leqslant n$ and in order to do that we need to extend the leading data to $\Bbbk[x]_>$ and switch from the notions of GB and NF to SB and weak NF. Later on, in Section 4.4 we are going to do that in a bit more general (non-commutative) setting.

4.2 Non-commutative localization

Let us recall some theory about non-commutative localization (in this section we follow [82, Chapters 9 and 10] and [25]).

Definition 4.2.1. Recall that

1. a subset S in a ring R is called a multiplicative set if $S \cdot S \subseteq S, 1 \in S$ and $0 \notin S$.
2. a homomorphism $\alpha : R \to R'$ is called **S-inverting** if all $\alpha(s), s \in S$ are units in R'.

4.2.1 Universal construction

Proposition 4.2.2 (Proposition (9.2) in [82]). *Let R be any ring with a multiplicative set $S \subseteq R$. Then there exists an S-inverting homomorphism ε from R to some ring, denoted by R_S, with the following universal property: For any S-inverting homomorphism $\alpha : R \to R'$, there exists a unique ring homomorphism $f : R_S \to R'$ such that $\alpha = f \circ \varepsilon$, that is, the following diagram commutes:*

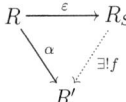

The universal property 4.2.2 guarantees the **uniqueness** of $\varepsilon : R \to R_S$ and therefore justifies the notation R_S for the receiving ring of the universal S-inverting homomorphism ε. Moreover the ring R_S is also called **universal S-inverting ring**.

Remark 4.2.3. The universal S-inverting homomorphism is injective iff R can be embedded into a ring in which all elements of S have inverses.

Remark 4.2.4. In contrary to the commutative case R_S is difficult to work with since generally universal S-inverting homomorphism does not have good properties known from the commutative case (4.2.5 b and c). Thus, for instance its elements have complicated forms (see equation (9.4) in [82]). Furthermore, the nature of R_S is rather unpredictable:

- R_S may be zero even if $R \neq 0$ (see example (9.3) in [82]).
- R_S may not be a domain even if R is a domain.
- $\varepsilon : R \to R_S$ may not be injective even if R is a domain and $S = R \setminus \{0\}$.

4.2.2 Ore construction

The following definition is modeled on the commutative case.

Definition 4.2.5. A ring R' is said to be a **right ring of fractions** (with respect to $S \subseteq R$) if there is a given ring homomorphism $\varphi : R \to R'$ such that:

(a) φ is S-inverting,

(b) every element of R' has the form $\varphi(a)\varphi(s)^{-1}$ for some $a \in R$ and $s \in S$,

(c) $\mathrm{Ker}(\varphi) = \{r \in R : rs = 0 \text{ for some } s \in S\}$.

Remark 4.2.6. In contrary to the general situation with R_S, due to (c) from the above, we always have $R' \neq 0$ if it exists.

4.2. NON-COMMUTATIVE LOCALIZATION

The set S is said to be a **right Ore set**, or **right permutable** if for any $a \in R$ and $s \in S, aS \cap sR \neq \emptyset$. The set S is said to be **right reversible** if for $a \in R$, if $s'a = 0$ for some $s' \in S$, then $as = 0$ for some $s \in S$. If the multiplicative set $S \subseteq R$ is both right permutable and right reversible, we shall say that S is a **right denominator set**.

There are, of course, all the corresponding "left" notions.

Theorem 4.2.7 (Theorem (10.6) in [82]). *The ring R has a right ring of fractions with respect to S (denoted by RS^{-1}) iff S is a right denominator set.*

Remark 4.2.8. In the proof of Theorem 4.2.7, the right ring of fractions of R with respect to S is constructed as $R \times S$ modulo the following relation:

$(a, s) \sim (a', s')$ (in $R \times S$) iff there exist $b, b' \in R$ such that $sb = s'b' \in S$ and $ab = a'b' \in R$. In what follows we will denote the equivalence class of (a, s) modulo $\sim$ by a/s.

Note that if we let $b' = 1$ in above, we get that $(a, s) \sim (ab, sb)$ as long as $sb \in S$. Therefore we can think of "$\sim$" as the best equivalence relation which "identifies" (a, s) with (ab, sb) ($\forall a \in R, s \in S, sb \in S$). Indeed, this generalizes the equivalence relations in the commutative case (cf. Section 4.1).

The binary operations on RS^{-1} are defined as follows: let $\frac{a}{s}, \frac{a'}{s'} \in RS^{-1}$ then since $s'S \cap sR \neq \emptyset$ there exist $r \in R, r' \in S$ such that $s'r' = sr \in S$, so that $\frac{a}{s} = \frac{ar}{sr}$ and $\frac{a'}{s'} = \frac{a'r'}{s'r'}$. Now

$$\frac{a}{s} + \frac{a'}{s'} := \frac{ar + a'r'}{t}, \text{ where } t = sr = s'r'.$$

Analogously, since $sR \cap a'S \neq \emptyset$ there exist $r \in R, r' \in S$ such that $a'r' = sr$. Now

$$\frac{a}{s} \cdot \frac{a'}{s'} := \frac{ar}{s'r'}.$$

The zero element in RS^{-1} is $\frac{0}{1}$ and the multiplicative identity is $\frac{1}{1}$.

Moreover, the defining homomorphism $\varphi : R \to RS^{-1}$ from definition 4.2.5 is given by $a \mapsto \frac{a}{1}$.

Corollary 4.2.9. *If S is a right denominator set, then $\varphi : R \to RS^{-1}$ is a universal S-inverting homomorphism. In particular, there is a unique isomorphism $g : R_S \to RS^{-1}$ such that $g \circ \varepsilon = \varphi$, where $\varepsilon : R \to R_S$, that is, the following diagram commutes:*

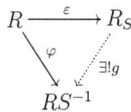

Remark 4.2.10. Let C be the center of R. If the set $S \subseteq C$ is a multiplicative set, then S is clearly both left and right denominator set, and we can identify $S^{-1}R$ with RS^{-1}. We speak of $RS^{-1} = S^{-1}R \cong R_S$ as a "central localization" of R.

Moreover, in this case the construction of RS^{-1} can be simplified since for any $a \in R, s \in S$ we have that $as = sa \in aS \cap sR$. Thus for any $\frac{a}{s}, \frac{a'}{s'} \in RS^{-1}$ we get that $a's = sa' \in sR \cap a'S$ and $t = ss' = s's \in s'S \cap sR$ so that $\frac{a}{s} = \frac{as'}{t}$ and $\frac{a'}{s'} = \frac{a's}{t}$. Therefore the operations are actually defined as in the usual commutative case:

$$\frac{a}{s} + \frac{a'}{s'} := \frac{as' + a's}{ss'}, \tag{4.1}$$

$$\frac{a}{s} \cdot \frac{a'}{s'} := \frac{aa'}{ss'}. \tag{4.2}$$

Remark 4.2.11. For commutative rings the above-mentioned "central localization" and the commutative localization coincide due to their universal properties.

Remark 4.2.12. The universal S-inverting homomorphism is injective iff S contains only non-zero-divisors.

Remark 4.2.13. We can apply localization to modules as well: let $\lambda : R \to R_S$ be the universal S-inverting homomorphism. Then to each left (resp. right) R-module M there corresponds a left (resp. right) R_S-module M_S with an R-module homomorphism $\mu : M \to M_S$ (where M_S is regarded as R-module by means of $\lambda : ax = \lambda(a)x$, resp. by $xa = x\lambda(a)$ for $x \in M_S, a \in R$) and μ is the universal mapping with this property, i.e. for any R-module homomorphism M into an R_S-module N, there exists a unique R_S-module homomorphism $g : M_S \to N$ such that the following diagram commutes:

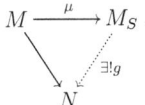

Moreover there is a formula for M_S in terms of tensor product (cf. e.g., [90]):

$$M_S = M \otimes_R R_S,$$

and if S is a right denominator set in R then $M_S = \left\{ \frac{m}{s} \mid m \in M, s \in S \right\} / \sim$, where $\sim$ is the usual equivalence relation on fractions.

4.3 Central localization

Let A be a commutative $\Bbbk$-algebra, $S \subset A$ be a multiplicative set (containing no zero-divisors) and G be any $\Bbbk$-algebra. Due to Remark 4.2.11 there exists a unique localization A_S.

Due to 4.2.12 the S-inverting homomorphism $\pi : A \to A_S$ is injective.

Moreover, it follows that the map $\pi \otimes_{\Bbbk} id_G : A \otimes_{\Bbbk} G \to A_S \otimes_{\Bbbk} G$ is injective as well.

4.3. CENTRAL LOCALIZATION

The set $S' := S \otimes_k 1_G = \{s \otimes_k 1_G \mid s \in S\} \subset A \otimes_k G$ is a multiplicative set in $A \otimes_k G$ which contains no zero-divisors, since S does so.

Since S' is central we may consider the "central localization" $(A \otimes_k G)_{S'}$.

Furthermore, due to 4.2.12 the S'-inverting homomorphism $\varepsilon : A \otimes_k G \to (A \otimes_k G)_{S'}$ is injective.

Let us picture all related maps on the following commutative diagram:

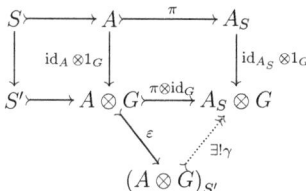

where tensor products are taken over the ground field k, the maps π, ε are localization maps and γ is a unique homomorphism due to universal property of $(A \otimes G)_{S'}$, since $\pi \otimes \mathrm{id}_G : A \otimes G \to A_S \otimes G$ is an S'-inverting homomorphism.

Proposition 4.3.1. *In the above diagramm the map γ is bijective.*

Proof. Since γ is a homomorphism and due to the universal property:

$$\gamma\left(\frac{a \otimes h}{s \otimes 1}\right) = \gamma\left((a \otimes h) \cdot (s \otimes 1)^{-1}\right) = \left(\frac{a}{1} \otimes h\right) \cdot \left(\frac{1}{s} \otimes 1\right) = \frac{a}{s} \otimes h.$$

The map γ is surjective since:

$$A_S \otimes_k G \ni \sum_i \frac{a_i}{s_i} \otimes g_i = \sum_i \frac{\hat{a}_i}{s} \otimes g_i = \gamma\left(\frac{\sum_i \hat{a}_i \otimes g_i}{s \otimes 1}\right),$$

where $s = \prod_i s_i, \hat{a}_i = a_i \cdot \prod_{j \neq i} s_j$. Note that all these sums and products are finite.

In order to see that γ is injective let us consider any $\frac{\sum_i a_i \otimes g_i}{s \otimes 1} \in (A \otimes_k G)_{S'} \cap \mathrm{Ker}(\gamma)$:

$$0 = \gamma\left(\frac{\sum_i a_i \otimes g_i}{s \otimes 1}\right) = \sum_i \frac{a_i}{s} \otimes g_i = \left(\frac{1}{s} \otimes 1\right)\left(\sum_i \frac{a_i}{1} \otimes g_i\right).$$

Since S has no zero-divisors it follows that $\left(\sum_i \frac{a_i}{1} \otimes g_i\right) = 0$ in $A_S \otimes_k G$ and thus $\left(\frac{1 \otimes 1}{s \otimes 1}\right)\left(\frac{\sum_i a_i \otimes g_i}{1 \otimes 1}\right) = 0$ in $(A \otimes_k G)_{S'}$. ∎

Thus instead of computing in $A_S \otimes G$ we will compute in $(A \otimes G)_{S'}$. The later resembles (for some localizing sets S) a ring which may be represented in SINGULAR as a ring associated to a monomial ordering (cf. [67]).

In the following we will check under which conditions on G we can apply this theory.

4.4 Rings Associated to Monomial Orderings: non-commutative setting

Let us introduce the notion of rings associated to monomial orderings in our non-commutative setting.

Let $A = \Bbbk[x_1, \ldots, x_n]$, $>$ be any monomial ordering on A. Let G be a Noetherian $\Bbbk$-algebra generated by variables $y_1, \ldots, y_m$, $>_G$ a global ordering on $\operatorname{Mon}(y_1, \ldots, y_m)$. We are mainly interested in the case of an exterior algebra but we also show how to deal with any G–algebra.

Let us consider the algebra

$$B := A \otimes_\Bbbk G$$

as a $\Bbbk$-algebra generated by variables $y_1, \ldots, y_m, x_1, \ldots, x_n$ together with the block ordering $>' = (>_G, >_A)$ (i.e. $\forall i, j : y_i >' x_j$) on $\operatorname{Mon}(y_1, \ldots, y_m, x_1, \ldots, x_n)$.

The set of commutative polynomials in A with constant leading term $S := \{u \in A \setminus \{0\} \mid \operatorname{Lm}(u) = 1\} \subset A$ is multiplicative. Denote

$$A_> = S^{-1}A =: \mathcal{A}.$$

Denote $S' := \{u \in B \setminus \{0\} \mid \operatorname{Lm}(u) = 1\} \subset B = A \otimes_\Bbbk G$. Clearly S' is multiplicatively closed and belongs to the center of B.

Moreover, S' equals to the image of S in B under the map $\operatorname{id}_A \otimes 1_G : A \to B$ since the ordering on B is the block ordering $>' = (>_G, >_A)$ and $>_G$ is global.

Definition 4.4.1. Due to Proposition 4.3.1 we get:

$$\mathcal{A} \otimes_\Bbbk G = (S^{-1}A) \otimes_\Bbbk G \cong S'^{-1}(A \otimes_\Bbbk G) = B_{>'} =: \mathcal{B}$$

and call it *the localization of B w.r.t. $>'$* (or $>$).

We intend to simulate $\mathcal{A} \otimes_\Bbbk G$ by working with polynomials from B with respect to the mixed monomial ordering $>'$. For this we need analogs of *standard basis* and *weak polynomial normal form* (from [67, Section 1.6]). Let us reintroduce them in our (non-commutative) setting. First of all we extend leading data to $\mathcal{B}$:

Definition 4.4.2. For $f \in \mathcal{B}$ choose $u \in B$ such that $\operatorname{Lt}(u) = 1$ (i.e. $u \in S'$) and $u \cdot f \in B$. We define $\operatorname{Lm}(f) := \operatorname{Lm}(uf), \operatorname{Lt}(f) := \operatorname{Lt}(uf)$ etc. Let $\mathcal{B}^*$ denote the set of units of $\mathcal{B}$.

4.4. RINGS ASSOCIATED TO MONOMIAL ORDERINGS

Note that the above definition is independent of the choice of u.

Definition 4.4.3. Let $I \subset B$ be a left (resp. right) ideal. A finite set $G \subset B$ is called a **left (resp. right) standard basis** of I if $G \subset I$ and for any $f \in I \setminus \{0\}$ there exists a $g \in G$ satisfying $\text{Lm}(g) | \text{Lm}(f)$ (in the usual sense).

Two-sided SBs are defined as before for G-algebras (cf. Remark 2.2.18).

Let us consider any non-zero ideal $I \subset B$. Since G is Noetherian (any GR-algebra is Noetherian, cf. Section 3.2 [86]) it follows that B is Noetherian as well. Therefore $\text{L}(I) \subset B$ is finitely generated. Thus we can choose a finite set of monomials $m_1, \ldots, m_s$ generating $\text{L}(I)$, by definition of the leading ideal these generators are leading monomials of suitable elements $g_1, \ldots, g_s \in I$. By definition, the set $\{g_1, \ldots, g_s\}$ is a standard basis for I. Hence every non-zero ideal in B has a standard basis.

Definition 4.4.4. A map

$$\text{NF} : \mathcal{B} \times \mathcal{G}_\mathcal{B} \to \mathcal{B}, (f, G) \mapsto \text{NF}(f \mid G),$$

is called a **left weak normal form** on $\mathcal{B}$ if, for all $G \in \mathcal{G}_\mathcal{B}, f \in \mathcal{B}$,

1. $\text{NF}(0 \mid G) = 0$,
2. $\text{NF}(f \mid G) \neq 0 \Rightarrow \text{Lm}(\text{NF}(f \mid G)) \notin \text{L}(G)$,
3. there exists a unit $u \in \mathcal{B}^*$ such that $uf - \text{NF}(f \mid G) \in {}_\mathcal{B}\langle G \rangle$, and if $G = \{g_1, \ldots, g_s\}, s \geqslant 0$ then $uf - \text{NF}(f \mid G)$ (or, by abuse of notation, uf) has a **standard representation** with respect to $\text{NF}(- \mid G)$, that is, there exists $a_i \in \mathcal{B}$ such that

$$uf - \text{NF}(f \mid G) = \sum_{i=1}^{s} a_i g_i, \qquad (4.3)$$

and

$$\text{Lm}\left(\sum_{i=1}^{s} a_i g_i\right) \geqslant \text{Lm}(a_i g_i), \quad \text{Lm}(a_i) *_\mathcal{B} \text{Lm}(g_i) \neq 0$$

for all i such that $a_i \neq 0$ and $g_i \neq 0$.

Definition 4.4.5. A weak normal form $\text{NF}(- \mid -)$ is called **polynomial** if, whenever $f \in B$ and $G \subset B$, there exists a unit $u \in \mathcal{B}^* \cap B$ such that uf has a standard representation with $a_i \in B$.

Definition 4.4.6. Let $f \in B \setminus \{0\}$ and $w = (w_1, \ldots, w_s)$ be a tuple of positive real numbers (*weights*), where s is the number of variables of B.

- The **ecart** of f is the integer $\text{Ecart}(f) := \text{Deg} f - \text{Deg} \text{Lm}(f)$.
- The **weighted ecart** is defined by $\text{Ecart}_w(f) := \text{w-Deg}(f) - \text{w-Deg}(\text{Lm}(f))$, where the weighted degree is defined by $\text{w-Deg}(\mathbf{y}^{\underline{\alpha}}) := \sum_{i=1}^{s} w_i \cdot \alpha_i$

Proposition 4.4.7. Let G be any G-algebra in variables $y_1, \ldots, y_n$. We can choose appropriate positive integer weights w_i for variables y_i and adjoin a commutative (fake) variable t with the weight $w_t = 1$ so that the non-commutative relations of the G-algebra $G^h = G \otimes_{\Bbbk} \Bbbk[t]$ become weighted homogeneous.

Proof. Let us consider all non-trivial non-commutative relations of G:

$$y_j * y_i = c_{ij} \cdot y_i y_j + d_{ij} =: f_{ij},$$

where $d_{ij} = \sum_k a_{ij}^k \cdot \mathbf{m_{ij}^k}$ and $y_i y_j >_G \mathbf{m_{ij}^k}$, $a_{ij}^k \in \Bbbk^*$, $\mathbf{m_{ij}^k} \in \text{Mon}(G)$.

Let M be the finite set of all monomials occurring in non-trivial non-commutative relations of G:

$$M = \{y_i y_j, \mathbf{m_{ij}^k}\} \subset \text{Mon}(G).$$

Due to the Lemma 1.2.11 from [67] there exists some weights $\underline{w} = (w_1, \ldots, w_n) \in \mathbb{Z}^n$ such that $\mathbf{m_1} >_G \mathbf{m_2}$ if and only if w-$\text{Deg}(\mathbf{m_1}) >$ w-$\text{Deg}_w(\mathbf{m_2})$ for all $\mathbf{m_1}, \mathbf{m_2} \in M$. Moreover, since $>_G$ is global $\underline{w}$ can be chosen such that all $w_i > 0$.

Denote $\delta_{ij}^k := $ w-$\text{Deg}(y_i y_j) - $ w-$\text{Deg}(\mathbf{m_{ij}^k})$. By construction δ_{ij}^k are non-negative integers. We homogenize all non-trivial f_{ij} w.r.t. w-Deg by multiplying its non-zero terms $a_{ij}^k \cdot \mathbf{m_{ij}^k}$ with appropriate powers of t as follows:

$$f_{ij}^h = c_{ij} \cdot y_i y_j + \sum_k a_{ij}^k \cdot t^{\delta_{ij}^k} \cdot \mathbf{m_{ij}^k}.$$

Thus we have got the following (homogenized) G-algebra

$$G^h := G \otimes_{\Bbbk} \Bbbk[t] = \Bbbk \langle y_1, \ldots, y_n, t \mid y_j * y_i = f_{ij}^h \rangle.$$

We endow G^h with the block ordering $(>_G, >_{nat})$.

Thus we have got a homogenization G^h of G together with the weighting $w^h := (w_1, \ldots, w_n, 1)$ with positive integers such that relations of G^h become homogeneous w.r.t. w-Deg_{w^h}.

Product of two homogeneous (w.r.t. w-Deg_{w^h}) polynomials from G^h remains homogeneous w.r.t. w-Deg_{w^h} due to homogeneous relations by induction on the number of non-commutative variables in G^h. ∎

Remark 4.4.8. If the original ordering on G is a degree-ordering then we don't need any weighting (all weights are equal to 1) but might still need a homogenization.

Examples 4.4.9. Let us consider several examples of non-commutative algebras:

1. **Weyl algebra**: $\Bbbk \langle x, D \mid D * x = xD + 1 \rangle$ endowed with any global ordering does not need weighting.

4.4. RINGS ASSOCIATED TO MONOMIAL ORDERINGS

2. **Heisenberg algebra**: $\Bbbk \langle x, y, h \mid y * x = xy + h^n \rangle$ is a G-algebra with respect to the lexicographical ordering with $x > y > h$ if $n > 2$ and requires weighting for homogenization, for instance: $w_h = 1, w_x = n - 1, w_y = 1$.

Remark 4.4.10. Clearly we do not need to homogenize relations in a graded commutative algebra. Thus a homogenization of a tensor product (as always over the field) of a G-algebra with a graded commutative algebra is just a homogenization of that G-algebra in the tensor product.

Algorithm 4.4.1 LEFTNFMORA(f, F)

ASSUME: B is a tensor product of a G-algebra with a graded commutative algebra.
INPUT: $f \in B, F \in \mathcal{G}_B$;
OUTPUT: $h \in B$, a polynomial left weak normal form of f w.r.t. F.
1: Choose w according to Proposition 4.4.7 and Remark 4.4.10;
2: $h := f$;
3: $T := F$;
4: **while** $((h \neq 0)$ and $(T_h := \{g \in T : \mathrm{Lm}(g) | \mathrm{Lm}(h)\} \neq \emptyset))$ **do**
5:   Choose $g \in T_h$ with $\mathrm{Ecart}_w(g)$ minimal;
6:   **if** $\mathrm{Ecart}_w(g) > \mathrm{Ecart}_w(h)$ **then**
7:     $T := T \cup \{h\}$;
8:   **end if**
9:   $h := \mathrm{LeftSPoly}(h, g)$;
10: **end while**
RETURN: h;

Proposition 4.4.11. *Let B be a tensor product of a G-algebra with a graded commutative algebra. Algorithm 4.4.1 terminates and computes a polynomial left weak normal form of $f \in B$ with respect to $F \in \mathcal{G}_B$.*

Proof. We are only going to verify that the proof for the commutative case (see Algorithm 1.7.6 in [67]) work in our non-commutative setting.

Correctness can be easily verified due to the following observations:

1. by construction $\mathrm{Lm}(g'_\nu)$ divides $\mathrm{Lm}(h_{\nu-1})$ and thus $\mathrm{Lm}(\mathrm{Lm}(g'_\nu) \cdot \mathrm{Lm}(m_\nu)) = \mathrm{Lm}(h_{\nu-1})$.
2. commutativity of multiplication was not employed in the original proof.
3. construction of $\mathrm{LeftSPoly}(h, g)$ is correct in any GR-algebra. Moreover, no leading term canceling due to zero divisors occurs since $\mathrm{Lm}(g)$ divides $\mathrm{Lm}(h)$.

For the termination we need to switch to the following homogenized version of the above algorithm, which works in B^h:
1: $h := f^h$;

2: $T := F^h := \{g^h : g \in F\}$;
3: **while** $((h \neq 0)$ and $(T_h := \{g \in T : \text{Lm}(g) \mid \tau^\alpha \text{Lm}(h) \text{ for some } \alpha\} \neq \emptyset))$ **do**
4:    Choose $g \in T_h$ with $\alpha \geqslant 0$ minimal;
5:    **if** $\alpha > 0$ **then**
6:       $T := T \cup \{h\}$;
7:    **end if**
8:    $h := \text{LeftSPoly}(\tau^\alpha h, g)$;
9:    Divide out maximal possible power of homogenizing variable τ from h;
10: **end while**
RETURN: $h|_B$;

The homogenizations of input data f^h, g^h in above are with respect to additional homogenizing commutative variable τ with unit weight exactly as in proposition 4.4.7 (so that all terms of f^h, g^h become of the same weighted degree).

Denote $\Bbbk[\tau] \otimes_\Bbbk B^h$ by $\widetilde{B}$. We endow $\text{Mon}\left(\widetilde{B}\right) = \text{Mon}(\tau, y_1, \ldots, y_m, t, x_1, \ldots, x_n)$ with the ordering $>_h$ given by $\tau^\alpha m_1 >_h \tau^\beta m_2$ if w-$\text{Deg}(\tau^\alpha m_1) >$ w-$\text{Deg}(\tau^\beta m_2)$ or if w-$\text{Deg}(\tau^\alpha m_1) =$ w-$\text{Deg}(\tau^\beta m_2)$ and $m_1 >_{B'} m_2$.

This defines a well-ordering on $\text{Mon}\left(\widetilde{B}\right)$ and we have

$$\text{Lm}_{>_h}\left(f^h\right) = \tau^{\text{Ecart}_w(f)} \text{Lm}_{>_{B^h}}(f).$$

In particular, $\text{Ecart}(f) = \text{Deg}_\tau \text{Lm}_{>_h}\left(f^h\right)$.

Note that h remain weighted homogeneous (in left S-polynomial) due to our homogenization.

Since $\widetilde{B}$ is Noetherian, and $>_h$ is a well-ordering the algorithm terminates due to the original argument.

∎

Remark 4.4.12. We can extend to the above weak normal form algorithm 4.4.1 to work with (homogeneous) vectors and modules, homogenized by assigning weights to canonical (free) module generators.

Remark 4.4.13. Clearly we do not need to homogenize relations in a graded commutative algebra and it is easy to see that the above weak polynomial normal form algorithm 4.4.1 is correct and terminates for graded commutative algebras without any homogenization.

Moreover, note that for polynomial graded homogeneous input the usual left normal form algorithm 3.3.1 terminates as well.

Remark 4.4.14. Let $\mathcal{A}$ be a graded commutative algebra endowed with a mixed ordering, such that anti-commuting variables are global. Let $\mathcal{R}$ be the corresponding central localization $\mathcal{A}_>$. Then a left SB of an ideal in $\mathcal{R}$, given by elements from $\mathcal{A}$ can be computed (due to the above and following [67]) using the BBA 3.4.1 with the normal form defined by Algorithm 4.4.1.

Chapter 5
Syzygies and free resolutions

Closely following [67] we generalize our theory for ideals, developed before to the case of free modules.

Furthermore we establish relations between Gröbner bases and syzygies over central localizations of graded commutative algebras. A starting point for this is the idea that syzygies of leading terms (w.r.t. some special ordering) give rise to syzygies of whole elements, which also leads to syzygy-driven Buchberger's algorithms (cf. [113, 114, 96]). This idea is formalized in Theorem 5.3.6, which is one of the main results of this thesis. This theorem may be considered as a generalization of the Buchberger's Criterion to the case of a central localization of a graded commutative algebra and, in particular, it also proves our characterization of a GB (resp. SB) in a (resp. central localization of a) graded commutative a algebra.

Following the approach due to [79, 81] we propose Algorithm 5.4.4 for computing free resolutions over central localizations of graded commutative algebras. For this we have introduced the notation of "elementary arrow", which allows us to avoid tedious business of keeping track of indices and ordered sets. It also makes it possible to formulate our algorithm very close to the implementation side (see also Section 6.5).

5.1 Computer Algebra for modules

Here we will generalize the computational approach from previous chapters to submodules of free modules over k-algebras.

Let from now on $\mathcal{A}$ stand for a graded commutative algebra endowed with a monomial ordering $>$ on the set of its monomials $\text{Mon}(\mathcal{A})$. We denote its odd (resp. even) variables by $\xi_1, \ldots, \xi_m$ (resp. $x_1, \ldots, x_n$).

Let us recall Section 1.4. Now, following [67], we extend the notion of monomial and monomial ordering to the free left $\mathcal{A}$-module $\mathcal{A}^r := \oplus_{i=1}^r \mathcal{A}\epsilon_i$, where

$$\epsilon_i = (0, \ldots, 0, 1_i, 0, \ldots, 0)^t \in \mathcal{A}^r$$

is the i-th canonical basis vector of $\mathcal{A}^r$. For any $m \in \mathrm{Mon}(\mathcal{A})$, we denote

$$m\epsilon_i := (0, \ldots, 0, m_i, 0, \ldots, 0)^t \in \mathcal{A}^r$$

and call it a **module monomial** (involving component i) or simply a **monomial**. Therefore the set of monomials of $\mathcal{A}^r$ looks as follows:

$$\mathrm{Mon}(\mathcal{A}^r) := \cup_{i=1}^r \mathrm{Mon}(\mathcal{A})\epsilon_i$$

In our adopted notations, a monomial multiplied with a coefficient is called a term. Therefore, we shall call $cm\epsilon_i$ a **module term** or simply a **term**, where $c \in \Bbbk, m \in \mathrm{Mon}(\mathcal{A}), 1 \leqslant i \leqslant r$.

Definition 5.1.1. We say that monomial $m_1\epsilon_i$ divides monomial $m_2\epsilon_j$ iff m_1 divides m_2 and $i = j$, in such a case the quotient of such monomials is defined by: $m_2\epsilon_i/m_1\epsilon_i := m_2/m_1 \in \mathrm{Mon}(\mathcal{A})$. Also we may say that $m_1 \in \mathrm{Mon}(\mathcal{A})$ divides $m_2\epsilon_j$ iff m_1 divides m_2, and in such a case $m_2\epsilon_j/m_1 := (m_2/m_1)\epsilon_j \in \mathrm{Mon}(\mathcal{A}^r)$.

Moreover, we denote

$$\mathrm{lcm}(m_1\epsilon_i, m_2\epsilon_j) := \begin{cases} \mathrm{lcm}(m_1, m_2)\epsilon_i, & \text{if } i = j \\ 0, & \text{otherwise.} \end{cases}$$

We extend the notion of signed quotient to module monomials as follows.

Definition 5.1.2. Let $A, B \in \mathrm{Mon}(\mathcal{A}^r)$ such that B divides A. Let us define a **signed quotient** of A and B by

$$(A /\!/ B) := \sigma C,$$

where $C = A/B \in \mathrm{Mon}(\mathcal{A}), \sigma = \mathrm{Lc}(C \cdot B) \in \{\pm 1\}$.

Moreover, we put $(0 /\!/ B) := 0$, for any (module) monomial B.

Remark 5.1.3. It is easy to see that $(A /\!/ B) \cdot B = A$, whereas $(A/B) \cdot B = \pm A$. Moreover, if $A \cdot B = \sigma C$, where $A \in \mathrm{Mon}(\mathcal{A}), B, C \in \mathrm{Mon}(\mathcal{A}^r), \sigma \in \{\pm 1\}$ then

$$A = \sigma \left(C /\!/ B \right) \tag{5.1}$$

Furthermore, if $A, B \in \mathrm{Mon}(\mathcal{A}^r)$ then for any $D \in \mathrm{Mon}(\mathcal{A}^r)$, which is divisible by B and divides A the following equality holds:

$$(A /\!/ B) = (A /\!/ D) \cdot (D /\!/ B) \tag{5.2}$$

5.1. COMPUTER ALGEBRA FOR MODULES

Definition 5.1.4. Let $>$ be a monomial ordering on $\text{Mon}(\mathcal{A})$. A **module monomial ordering** or simply a **monomial ordering** on $\mathcal{A}^r$ is a total ordering $\succ$ on $\text{Mon}(\mathcal{A}^r)$ satisfying the following conditions:

$$m_1 > m_2 \Longrightarrow m_1\epsilon_i \succ m_2\epsilon_i, \tag{5.3}$$
$$m_1\epsilon_i \succ m_2\epsilon_j \Longrightarrow \text{Lm}(m \cdot m_1)\epsilon_i \succ \text{Lm}(m \cdot m_2)\epsilon_j, \tag{5.4}$$

for all $m, m_1, m_2 \in \text{Mon}(\mathcal{A}), i, j = 1, \ldots, n$, whenever $m \cdot m_1 \neq 0$ and $m \cdot m_2 \neq 0$.

Note that, each component of $\mathcal{A}^r$ carries the ordering of $\mathcal{A}$ due to condition (5.3), which, in particular, implies that, the module ordering $\succ$ is a well-ordering iff the monomial ordering $>$ is a well-ordering. We call the module ordering $\succ$ **global** (resp. **local**, resp **mixed**) whenever $>$ is global (resp. local, resp mixed).

Examples 5.1.5. Standard examples of monomial orderings of a particular interest are the following:

term over position ordering, denoted by $(>, c)$, is defined by:

$$m_1\epsilon_i \succ m_2\epsilon_j \Leftrightarrow \begin{cases} m_1 < m_2 \text{ or} \\ m_1 = m_2 \text{ and } i < j, \end{cases}$$

position over term ordering, denoted by $(c, >)$, is defined by:

$$m_1\epsilon_i \succ m_2\epsilon_j \Leftrightarrow \begin{cases} i < j \text{ or} \\ i = j \text{ and } m_1 > m_2, \end{cases}$$

From now on, let us fix a module ordering $\succ$ and denote it also $>$.

Definition 5.1.6. Since any non-zero vector $f \in \mathcal{A}^r$ can be uniquely written as $f = cm\epsilon_i + f'$, with $c \in \Bbbk \setminus \{0\}$ and $m\epsilon_i > m'\epsilon_j$ for any non-zero term $c'm'\epsilon_j$ of f' (where $m, m' \in \text{Mon}(\mathcal{A})$) we may denote as usual $\text{Lm}(f) := m\epsilon_i, \text{Lc}(f) := c, \text{Lt}(f) := cm\epsilon_i$, and call them the **leading monomial**, **leading coefficient** and **leading term**, respectively, of f.

As in the ideal (polynomial) case, let us denote the set of module monomials occurring in f with non-zero coefficients by $\text{Mon}(f)$.

For any subset $F \subset \mathcal{A}^r$, we call

$$\text{L}(F) := {}_{\mathcal{A}}\langle \text{Lm}(f) \mid f \in F \setminus \{0\}\rangle \subset \mathcal{A}^r,$$

the **leading module** of F.

Note that the leading module has all the properties of leading ideal shown in Section 3.3.

Remark 5.1.7. Since $\mathrm{Mon}(\mathcal{A}^r)$ can be identified with a subset of $\mathbb{N}^N$, for some $N \in \mathbb{N}$ and using Dickson's Lemma (cf. 2.1.10) for $\mathbb{N}^N$ we conclude that any monomial submodule of $\mathcal{A}^r$ (for any r) is finitely generated. That is why we always can choose a minimal generating system for any leading (sub)-module.

Recall that for a fixed ordering $>$ on $\mathrm{Mon}(\mathcal{A})$ we denoted the associated localization of $\mathcal{A}$ w.r.t. $>$ (cf. Chapter 4) by
$$\mathcal{R} := \mathcal{A}_> := S_>^{-1}\mathcal{A}.$$

Our theory of SBs for ideals can be carried out to modules almost without any changes. In what follows we are going to formulate the relevant definitions and theorems while omitting the proofs, which are identical to the ideal case.

After fixing a module ordering on $\mathrm{Mon}(\mathcal{R}^r)$, as for ideals, we define:

Definition 5.1.8. 1. Let $M \subset \mathcal{R}^r$ be a submodule. A finite set $F \subset M$ is called a **standard basis** of M iff $\mathrm{L}(M) = \mathrm{L}(F)$, that is, for any nonzero $g \in M$ there exists an $f \in F$ such that $\mathrm{Lm}(g)$ is divisible by $\mathrm{Lm}(f)$.
2. In the case of a well-ordering, a standard basis F is called a **Gröbner basis**. In this case $\mathcal{R} = \mathcal{A}$ and therefore, $F \subset M \subset \mathcal{A}^r$.
3. A set $F \subset \mathcal{R}^r$ is called **interreduced** (or **irredundant**) if $0 \notin F$ and for each $f \in F$ holds $\mathrm{Lm}(f) \notin \mathrm{L}(F \setminus \{f\})$. An interreduced standard basis is also called **minimal** since in such a case the leading module $\mathrm{L}(_\mathcal{R}\langle F\rangle)$ is minimally generated by monomials $\mathrm{Lm}(f), f \in F$.

Definition 5.1.9. Let $\mathcal{G}$ denote the set of all finite ordered subsets of $\mathcal{R}^r$.

1. A map
$$\mathrm{NF}: \mathcal{R}^r \times \mathcal{G} \to \mathcal{R}^r, (f, G) \mapsto \mathrm{NF}(f \mid G),$$
is called a **left normal form**normal form on $\mathcal{R}^r$ if, for all $G \in \mathcal{G}, f \in \mathcal{R}^r$,
   (a) $\mathrm{NF}(0 \mid G) = 0$,
   (b) $\mathrm{NF}(f \mid G) \neq 0 \Rightarrow \mathrm{Lm}(\mathrm{NF}(f \mid G)) \notin \mathrm{L}(G)$,
   (c) $f - \mathrm{NF}(f \mid G) \in {}_\mathcal{A}\langle G\rangle$, and if $G = \{g_1, \ldots, g_s\}$ then $f - \mathrm{NF}(f \mid G)$ (or, by abuse of notation, f) has a **standard representation** with respect to $\mathrm{NF}(- \mid G)$, that is,
$$f - \mathrm{NF}(f \mid G) = \sum_{i=1}^{s} a_i g_i, \ a_i \in \mathcal{R}, \tag{5.5}$$
satisfying $\mathrm{Lm}(\sum_{i=1}^s a_i g_i) \geqslant \mathrm{Lm}(a_i g_i)$ and $\mathrm{Lm}(a_i) \cdot \mathrm{Lm}(g_i) \neq 0$ for all i such that $a_i g_i \neq 0$.

2. NF is called a **weak normal form** if instead of condition 1c only the following condition holds:
   (c') for each $f \in \mathcal{R}^r$ and each G there exists a commutative unit $u \in \mathcal{R}$ such that uf has a standard representation w.r.t. $\mathrm{NF}(- \mid G)$.

5.1. COMPUTER ALGEBRA FOR MODULES

3. NF is called **polynomial** if, for $f \in \mathcal{A}^r, G \subset \mathcal{A}^r$, then $u, a_i \in \mathcal{A}$.

Lemma 5.1.10. *Let $M \subset \mathcal{R}^r$ be a submodule, $F \subset M$ a SB of M and $\mathrm{NF}(- \mid F)$ a weak normal form on $\mathcal{R}^r$ wrt F.*

1. *For any $f \in \mathcal{R}^r$ we have that $f \in M$ iff $\mathrm{NF}(f \mid F) = 0$.*
2. *If $M' \subset \mathcal{R}^r$ is a submodule with $M \subset M'$, then from $\mathrm{L}(M') = \mathrm{L}(M)$ follows that $M = M'$.*
3. $M = {}_{\mathcal{A}}\langle F \rangle$, *that is, F generates M as an $\mathcal{R}$-module.*

Definition 5.1.11. *The **(left) S-polynomial** between vectors $f, g \in \mathcal{R}^r$ is defined by:*

$$\mathrm{LeftSPoly}(f, g) := \mathrm{Lc}(g)\,(m /\!/ \mathrm{Lm}(f)) \cdot f - \mathrm{Lc}(f)\,(m /\!/ \mathrm{Lm}(g)) \cdot g \in \mathcal{R}^r,$$

where $m := \mathrm{lcm}(\mathrm{Lm}(f), \mathrm{Lm}(g)) \in \mathrm{Mon}(\mathcal{A}^r) \cup \{0\}$.

Clearly $\mathrm{LeftSPoly}(f, g)$ is zero if leading terms have different components.

Definition 5.1.12. *Assuming that $\mathrm{w}\text{-}\mathrm{Deg}(-)$ is a (weighted) degree function on $\mathrm{Mon}(\mathcal{A})$ we extend it to all of $\mathrm{Mon}(\mathcal{A}^r)$ by weighting module-components with the weights $e_1 \ldots, e_r \in \mathbb{R}$ and putting:*

$$\mathrm{w}\text{-}\mathrm{Deg}(m\boldsymbol{\epsilon}_i) := \mathrm{w}\text{-}\mathrm{Deg}(m) + e_i,$$

for any monomial $m\boldsymbol{\epsilon}_i \in \mathrm{Mon}(\mathcal{A}^r)$.

For $f \in \mathcal{A}^r \setminus \{0\}$, let $\mathrm{w}\text{-}\mathrm{Deg}(f)$ be the maximal degree of all monomials occurring in f.

*Define the **ecart** of f as*

$$\mathrm{Ecart}(f) := \mathrm{w}\text{-}\mathrm{Deg}(f) - \mathrm{w}\text{-}\mathrm{Deg}(\mathrm{Lm}(f)).$$

The Algorithms LEFTNF (3.3.1), LEFTNFMORA (4.4.1), LEFTREDNF (2.2.1) and LEFTGB (3.4.1) carry over verbatim to the module case by replacing $\mathcal{A}$ to $\mathcal{A}^r$ (and $\mathcal{R}$ to $\mathcal{R}^r$) due to the following theorem:

Theorem 5.1.13. *Let $M \subset \mathcal{R}^r$ be a submodule and $F = \{f_1, \ldots, f_s\} \subset M$. Let $\mathrm{NF}(- \mid F)$ be a weak normal form on $\mathcal{R}^r$ w.r.t. F. Then the following are equivalent:*

1. *F is a SB of M.*
2. *$\mathrm{NF}(g \mid F) = 0$ for all $g \in M$.*
3. *Each $g \in M$ has a standard representation w.r.t. $\mathrm{NF}(- \mid F)$.*
4. *F generates M and $\forall g \in \mathrm{S}(F)\ \exists F_g \subset F : \mathrm{NF}(g \mid F_g) = 0$, where*

$$\mathrm{S}(F) := \{\mathrm{LeftSPoly}(f_i, f_j)\}_{1 \leqslant i,j \leqslant s} \bigcup \{\xi_l \cdot f_i \mid \xi_l \cdot \mathrm{Lm}(f_i) = 0, 1 \leqslant l \leqslant m\}_{1 \leqslant i \leqslant s} \subset \mathcal{R}^r.$$

Proof. The proof of $1 \Leftrightarrow 2 \Leftrightarrow 3 \Rightarrow 4$ is a similar to the ideal case, while the only remaining implication will be shown later in Theorem 5.3.6. ■

Computation of syzygy module (cf. Algorithm 5.1.1) can be considered as a byproduct of SB computation due to the following lemma:

Lemma 5.1.14. *Let $>_1$ be a module monomial ordering on $\bigcup_{i=1}^{r} \text{Mon}(\mathcal{A})\epsilon_i$. Let $M = {}_\mathcal{R}\langle f_1,\ldots,f_s\rangle \subset \mathcal{R}^r = \bigoplus_{i=1}^{r} \mathcal{R}\epsilon_i$. Consider the canonical embedding*

$$\mathcal{R}^r \subset \mathcal{R}^{r+s} = \bigoplus_{i=1}^{r+s} \mathcal{R}\epsilon_i,$$

and the canonical projection $\pi : \mathcal{R}^{r+s} \to \mathcal{R}^s$. Choose any module monomial ordering $>_2$ on $\bigcup_{i=r+1}^{r+s} \text{Mon}(\mathcal{A})\epsilon_i$. Construct the following "syzygy" ordering $>_{\text{Syz}}$ on $\text{Mon}(\mathcal{A}^{r+s})$:

$$m_1\epsilon_i >_{\text{Syz}} m_2\epsilon_j :\Leftrightarrow \begin{cases} m_1\epsilon_i >_1 m_2\epsilon_j, & \text{if } i \leqslant r \text{ and } j \leqslant r \\ m_1\epsilon_i >_2 m_2\epsilon_j, & \text{if } i \geqslant r \text{ and } j \geqslant r \\ i > j, & \text{otherwise} \end{cases}$$

Let G be a SB of $M' := {}_\mathcal{R}\langle f_1 + \epsilon_{r+1},\ldots,f_s + \epsilon_{r+s}\rangle$ w.r.t. $>_{\text{Syz}}$. Suppose that $G \cap \bigoplus_{i=r+1}^{r+s} \mathcal{R}\epsilon_i = \{g_1,\ldots,g_l\}$, then

$$\text{Syz}(f_1,\ldots,f_s) = \langle \pi(g_1),\ldots,\pi(g_l)\rangle.$$

Proof. The proof goes verbatim as in the commutative polynomial case (cf. Lemma 2.5.3 from [67]).

For algebras with zero-divisors a similar statement was also shown in [102][Propositions 7.3, 7.4]. ■

Algorithm 5.1.1 SYZ(F)

ASSUME: Let $>$ be an ordering on $\text{Mon}(\mathcal{A})$ and $\mathcal{R} = \mathcal{A}_>$
INPUT: $F = (f_1,\ldots,f_s) \in \mathcal{G}_{\mathcal{A}^r}$.
OUTPUT: $S \subset \mathcal{A}^s$, such that ${}_\mathcal{R}\langle S\rangle = \text{Syz}(f_1,\ldots,f_s) \subset \mathcal{R}^s$.
1: $F := \{f_1 + \epsilon_{r+1},\ldots,f_s + \epsilon_{r+s}\}$, where $\epsilon_1,\ldots,\epsilon_{r+s}$ denote the canonical generators of $\mathcal{R}^{r+s} = \mathcal{R}^r \oplus \mathcal{R}^s$ such that $f_1,\ldots,f_s \in \mathcal{R}^r = \bigoplus_{i=1}^{r} \mathcal{R}\epsilon_i$;
2: compute a SB G of $\langle F\rangle \subset \mathcal{R}^{r+s}$ w.r.t. $>_{\text{Syz}}$;
3: $G_0 := G \cap \bigoplus_{i=r+1}^{r+s} \mathcal{R}\epsilon_i = \{g_1,\ldots,g_l\}$, where $g_i = \sum_{j=1}^{s} a_{ij}\epsilon_{r+j}, i = 1,\ldots,l$;
RETURN: $\{(a_{i,1},\ldots,a_{i,s})\}_{i=1}^{l}$;

5.1. COMPUTER ALGEBRA FOR MODULES

Remark 5.1.15. Algorithm 5.1.1 is just a way to do bookkeeping, while computing a SB. Both input and output of Algorithm 5.1.1 may be interpreted as single matrices:

- the output $S \subset \mathcal{A}^s$ – as a matrix from $\text{Mat}_{s \times l}(\mathcal{A})$ consisting of elements a_{ij},
- the input $\{f_1, \ldots, f_s\} \subset \mathcal{A}^r$ – as a matrix from $\text{Mat}_{r \times s}(\mathcal{A})$, with columns being the component-wise representations of $f_1, \ldots, f_s$.

We will usually write the $\text{Syz}(f_1, \ldots, f_s)$ instead of $\text{Syz}(M)$ since due to the following proposition (5.1.16) the notion of syzygies of a module is only defined up to an isomorphism:

Proposition 5.1.16. *Let* $f_1, \ldots, f_s$ *and* $g_1, \ldots, g_k$ *are ordered sets of elements of* $\mathcal{R}^r$ *such that* $\langle f_1, \ldots, f_s \rangle = \langle g_1, \ldots, g_k \rangle = M$. *Then there exist free* $\mathcal{R}$*-modules* L *and* L' *such that*

$$\text{Syz}(f_1, \ldots, f_s) \oplus L \cong \text{Syz}(g_1, \ldots, g_k) \oplus L'.$$

Proof. This follows from the observation that:

$$\text{Syz}(f_1, \ldots, f_s, g_1, \ldots, g_k) = N \oplus L, N \cong \text{Syz}(f_1, \ldots, f_s).$$

This observation can be shown by considering a weak normal form NF on $\mathcal{R}^r$ and using standard representations of g_i w.r.t. $\text{NF}(- \mid f_1, \ldots, f_s)$. ∎

Successively computing syzygies of syzygies, we obtain an algorithm (5.1.2) to compute free resolutions up to any given length.

Remark 5.1.17. Unfortunately, as soon as $\mathcal{A}$ have zero-divisors one cannot expect any analogy for Hilbert's Syzygy Theorem. For instance, consider a free resolution of any odd variable:

$$0 \leftarrow {}_{\mathcal{A}}\langle \xi \rangle \xleftarrow{\xi} \mathcal{A} \xleftarrow{\xi} \mathcal{A} \xleftarrow{\xi} \cdots.$$

Therefore, we always need an additional termination condition for our free resolution algorithms (such as length).

Algorithm 5.1.2 RESOLUTION(F, l)

ASSUME: Let $>$ be an ordering on $\mathrm{Mon}(\mathcal{A})$ and $\mathcal{R} = \mathcal{A}_>$
INPUT: $F = (f_1, \ldots, f_s) \in \mathcal{G}_{\mathcal{A}^r}$, $l \in \mathbb{N}$
OUTPUT: A list of matrices $A_1, \ldots, A_l$ with $A_i \in \mathrm{Mat}_{s_{i-1} \times s_i}(\mathcal{A})$, $1 \leqslant i \leqslant l$, such that

$$\cdots \to \mathcal{R}^{s_l} \xrightarrow{\varphi_l} \mathcal{R}^{s_{l-1}} \to \cdots \to \mathcal{R}^{s_1} \xrightarrow{\varphi_1} \mathcal{R}^r \to \mathcal{R}^r/\langle f_1, \ldots, f_s \rangle \to 0,$$

is a free resolution of $\mathcal{R}^r/\langle f_1, \ldots, f_s \rangle$, where homomorphisms φ_i are defined by matrices A_i due to (1.3).
1: $i := 1, s_0 := r, s_1 := s$;
2: $A_1 := \mathrm{Matrix}(f_1, \ldots, f_s) \in \mathrm{Mat}_{r \times s}(\mathcal{A})$;
3: **while** $i < l$ **do**
4: $i := i + 1$;
5: $A_i := \mathrm{Syz}(A_{i-1})$; // Using Algorithm 5.1.2
6: **end while**
RETURN: $A_1, \ldots, A_l$.

5.2 Assumptions on orderings

In the case of a G-algebra a monomial ordering on power products is uniquely induced by a monoid total order on exponents. The requirements behind that is that power products correspond uniquely to exponents, which is not the case for quotient algebras. Fortunately, in our case power products correspond either to zero or to a unique exponent, such that degrees of odd variables are in $\{0, 1\}$. Moreover a module ordering must be compatible (cf. Definition 5.1.4).

Unfortunately, in order to introduce a Schreyer ordering on $\mathrm{Mon}(\mathcal{A}^s)$ induced by $X_1, \ldots, X_s \subset \mathrm{Mon}(\mathcal{A}^r)$ we need to compare sums of exponents, one comming from a power product and the other comming from a module monomial, since we cannot simply compare the corresponding products due to zero-divisors.

Up until now we were operating without any further assumptions on the ordering on monomials. Here we recall that the monomial ordering on $\mathrm{Mon}(\mathcal{A})$ is induced by a total ordering on exponents.

In this section we formalize our requirements/assumptions about the total ordering on the set of exponents .

Let the set of exponents

$$\Upsilon := \mathbb{N}^{n+m} \times \{0, 1, \ldots, r\}$$

be endowed with a total ordering, and

$$\mathrm{Exp} : \mathrm{Mon}(\mathcal{A}) \to \Upsilon, x^\alpha \mapsto \alpha \times \{0\}$$

5.2. ASSUMPTIONS ON ORDERINGS

be an injective map. Denote

$$\operatorname{Exp}(\mathcal{A}) := \{\operatorname{Exp}(m) \mid m \in \operatorname{Mon}(\mathcal{A})\} \subseteq \mathbb{N}^{n+m} \times \{0\},$$
$$\Upsilon^0 := \{\alpha \mid \alpha \times \{0\} \in \operatorname{Exp}(\mathcal{A})\} \subseteq \mathbb{N}^{n+m}.$$

Let us endow Υ with the following operation of $\operatorname{Exp}(\mathcal{A})$

$$\oplus : \operatorname{Exp}(\mathcal{A}) \times \Upsilon \to \Upsilon, (\alpha \times \{0\}, \beta \times \{i\}) \mapsto (\alpha + \beta) \times \{i\},$$

where $\alpha, \beta \in \mathbb{N}^k$, and $\alpha + \beta$ is the component-wise sum.

We require the following condition:

$$\alpha < \beta \implies \gamma \oplus \alpha < \gamma \oplus \beta, \forall \gamma \in \operatorname{Exp}(\mathcal{A}), \forall \alpha, \beta \in \Upsilon. \tag{5.6}$$

Definition 5.2.1. An *ordering* $>_1$ on $\operatorname{Mon}(\mathcal{A})$, *induced by* Exp is defined as follows: $m_1 >_1 m_2$ by definition iff $\operatorname{Exp}(m_1) > \operatorname{Exp}(m_2)$, where $m_1, m_2 \in \operatorname{Mon}(\mathcal{A})$.

Note that this induced ordering is total, and indeed a monomial ordering due to (5.6).

Recall that

$$\mathcal{A}^r = \bigoplus_{i=1}^{r} \mathcal{A}\epsilon_i, \quad \operatorname{Mon}(\mathcal{A}^r) := \{m\epsilon_i \mid m \in \operatorname{Mon}(\mathcal{A}), 1 \leqslant i \leqslant r\}$$

and extend the map Exp to $\operatorname{Mon}(\mathcal{A}^r)$ by putting

$$\operatorname{Exp} : x^\alpha \epsilon_i \mapsto \alpha \times \{i\} \in \Upsilon,$$

where $x^\alpha \in \operatorname{Mon}(\mathcal{A})$ and $i \in \mathbb{N}_r := \{1, \ldots, r\}$.

Definition 5.2.2. An *ordering* $>_1$ on $\operatorname{Mon}(\mathcal{A}^r)$, *induced by* Exp is defined as follows: $m_1 \epsilon_i >_1 m_2 \epsilon_j$ by definition iff $\operatorname{Exp}(m_1 \epsilon_i) > \operatorname{Exp}(m_2 \epsilon_j)$, where $m_1 \epsilon_i, m_2 \epsilon_j \in \operatorname{Mon}(\mathcal{A}^r)$.

Note that the induced ordering on $\operatorname{Mon}(\mathcal{A}^r)$ is total.

In order to be compatible with the usual definition of a module monomial ordering, an ordering on Υ must satisfy the following conditions:

1. $\alpha \times \{0\} < \beta \times \{0\} \implies \alpha \times \{i\} < \beta \times \{i\}$, for all $\alpha, \beta \in \Upsilon^0$ and $i \in \mathbb{N}_r$.
2. $\alpha \times \{i\} < \beta \times \{j\} \implies (\alpha + \gamma) \times \{i\} < (\beta + \gamma) \times \{j\}$, for all $\alpha, \beta, \gamma \in \Upsilon^0$ and $i, j \in \mathbb{N}_r$.

5.3 Schreyer ordering and syzygies of leading terms

Using our assumptions about exponents and following Definition 3.1 from [81] we generalize the notion of Schreyer ordering.

Recall that $\mathcal{A}$ denotes a graded commutative algebra endowed with an ordering $>$ and $\mathcal{R} = \mathcal{A}_>$.

Definition 5.3.1. Given $X_1, \ldots, X_s \in \mathrm{Mon}(\mathcal{A}^r)$. A module ordering $>_1$ on $\mathrm{Mon}(\mathcal{A}^s)$ is called **Schreyer ordering** (induced by $X_1, \ldots, X_s$) if it satisfies the following condition:

$$\mathrm{Exp}(m_1) \oplus \mathrm{Exp}(X_i) > \mathrm{Exp}(m_2) \oplus \mathrm{Exp}(X_j) \Longrightarrow m_1 \epsilon_i >_1 m_2 \epsilon_j, \quad (5.7)$$

for all $1 \leqslant i, j \leqslant s, m_1, m_2 \in \mathrm{Mon}(\mathcal{A})$ (i.e. for all $m_1 \epsilon_i, m_2 \epsilon_j \in \mathrm{Mon}(\mathcal{A}^s)$).

Remark 5.3.2. Clearly the condition (5.7) is not sufficient to define a total ordering on $\mathrm{Mon}(\mathcal{A}^s)$ but it can be refined to become one by an extra condition (say $\mathrm{P}(m_1, i, m2, j)$) to get a (total) monomial ordering:

$$m_1 \epsilon_i <_1 m_2 \epsilon_j \Leftrightarrow \begin{cases} \mathrm{Exp}(m_1) \oplus \mathrm{Exp}(X_i) < \mathrm{Exp}(m_2) \oplus \mathrm{Exp}(X_j) \text{ or} \\ \mathrm{Exp}(m_1) \oplus \mathrm{Exp}(X_i) = \mathrm{Exp}(m_2) \oplus \mathrm{Exp}(X_j) \text{ and } \mathrm{P}(m_1, i, m2, j), \end{cases} \quad (5.8)$$

where $1 \leqslant i, j \leqslant s, m_1, m_2 \in \mathrm{Mon}(\mathcal{A})$.

There are two orderings, which are called **Schreyer ordering**. Here we will call them "classical":

Definition 5.3.3. The **classical Schreyer ordering** is defined due to (5.8), where

$$\mathrm{P}(m_1, i, m2, j) := \text{"}i < j\text{"}. \quad (5.9)$$

The **reversed classical Schreyer ordering** is defined analogously with

$$\mathrm{P}(m_1, i, m2, j) := \text{"}i > j\text{"}. \quad (5.10)$$

Note that both classical Schreyer orderings from definition 5.3.3 are, in fact, module term orderings, whereas the *position over term* and *term over position* are, in general, not Schreyer orderings.

Recall that two (module) monomials $Y \mid X$ we denote $(X /\!/ Y) := \sigma \cdot (X/Y)$, for $\sigma \in \{\pm 1\} \subset \Bbbk^*$ such that $\sigma \cdot (X/Y) \cdot Y = X$. Moreover, we put $(0 /\!/ Y) := 0$ for any (module) monomial Y.

The following result, which seems to be new, is important for our algorithm to compute syzygies of modules over graded commutative algebras. Note that this result is ordering-independent.

5.3. SCHREYER ORDERING AND SYZYGIES OF LEADING TERMS

Proposition 5.3.4. *Let $\mathcal{A}$ be a graded commutative algebra with odd variables denoted by $\xi_1, \ldots, \xi_m$. Let $c_i \in \Bbbk^*, X_i \in \text{Mon}(\mathcal{A}^r), X_{ij} := \text{lcm}(X_i, X_j) \in \text{Mon}(\mathcal{A}^r), 1 \leqslant i, j \leqslant s$. Then the module $\text{Syz}(c_1 X_1, \ldots, c_s X_s)$ is generated by the following syzygies, which we shall call* **elementary syzygies***:*

$$\{\xi_l \epsilon_i : \xi_l \cdot X_i = 0, 1 \leqslant l \leqslant m\} \bigcup \{c_j (X_{ij}/\!\!/X_i) \epsilon_i - c_i (X_{ij}/\!\!/X_j) \epsilon_j : i < j\} \subseteq \mathcal{A}^s. \quad (5.11)$$

Proof. By construction and due to defining relations of $\mathcal{A}$ it is clear that

$$\{\xi_l \epsilon_i : \xi_l \mid X_i\} \bigcup \{c_j (X_{ij}/\!\!/X_i) \epsilon_i - c_i (X_{ij}/\!\!/X_j) \epsilon_j : 1 \leqslant i < j \leqslant s\} \subseteq \text{Syz}(c_1 X_1, \ldots, c_s X_s).$$

In order to prove the other inclusion let us consider nonzero $S := \sum_{i=1}^{s} h_i \epsilon_i \in \text{Syz}(c_1 X_1, \ldots, c_s X_s), h_i \in \mathcal{A}$, that is,

$$S * \{X_i\} = h_1 c_1 X_1 + \ldots + h_s c_s X_s = 0. \quad (5.12)$$

It is clear that for any monomial $m \in \text{Mon}(\mathcal{A}^r)$ its coefficient in (5.12) must be zero. Note that S is a finite sum of terms. Let

$$E := \text{Max}_{X \epsilon_j \in \text{Mon}(S)} \text{Exp}(X) + \text{Exp}(X_j),$$

$$S' := \sum_{c' X' \epsilon_j \text{ is a term of } S : \text{Exp}(X) + \text{Exp}(X_j) = E} c' X' \epsilon_j.$$

We consider the corresponding combination:

$$S' * \{x_i X_i\} = \sum_{\text{Exp}(X') + \text{Exp}(X_j) = E} c' c_j X' X_j \quad (5.13)$$

It must be zero. Hence by definition $S' \in \text{Syz}(c_1 X_1, \ldots, c_s X_s)$. Therefore by induction on the number or terms in S we only need to prove the statement for S'.

There are two cases: either all products in (5.13) $X' X_j$ are zeroes or they all are multiples of some monom $m \in \text{Mon}(\mathcal{A}^r)$ and cancel out all together.

For the former case, note that the product $X' X_j = 0$ iff there exists a ξ_l dividing both X' and X_j. Thus $c' X' \epsilon_j = c' (X'/\!\!/\xi_l) \xi_l \epsilon_j$ with $\xi_l \cdot X_j = 0$.

For the later case we need to proceed exactly as in the commutative polynomial case while taking case about signs. That is, we consider the case $S' = \sum_{i=1}^{s} b'_i Y_i \epsilon_i$, where w.l.o.g. we may assume that all $b'_i \neq 0$ and $Y_i X_i = \sigma_i m$, where $\sigma_i \in \{\pm 1\}$. Note that for all

$1 \leqslant i, j \leqslant s$: X_i and X_j divide X_{ij}, which in turn divides m. Denote $b_i := b'_i \sigma_i$. It is clear that
$$S' = \sum_{i=1}^{s} b'_i Y_i \epsilon_i = \sum_{i=1}^{s} b_i \left(m /\!/ X_i\right) \epsilon_i, \tag{5.14}$$
and that $(m /\!/ X_i) = (m /\!/ X_{ij})(X_{ij} /\!/ X_i)$.

Consider two summands as from the later sum in (5.14), $B, D \in \mathbb{k}^*$:

$$B (m /\!/ X_i) \epsilon_i + D (m /\!/ X_j) \epsilon_j =$$
$$\frac{Bc_j}{c_j} (m /\!/ X_{ij})(X_{ij} /\!/ X_i) \epsilon_i \pm \frac{Bc_i}{c_j} (m /\!/ X_{ij})(X_{ij} /\!/ X_j) \epsilon_j + D (m /\!/ X_j) \epsilon_j =$$
$$\frac{B}{c_j} (m /\!/ X_i) \cdot \underbrace{\left(c_j (X_{ij} /\!/ X_i) \epsilon_i - c_i (X_{ij} /\!/ X_j) \epsilon_j \right)}_{\text{needed elementary syzygy}} + \frac{c_i B + c_j D}{c_j} (m /\!/ X_j) \epsilon_j \quad (5.15)$$

Applying (5.15) repeatedly to (5.14) we will get $\frac{\sum_{i=1}^{s} c_i b_i}{c_s} (m /\!/ X_s) \epsilon_s$ as the remainder, which is zero since S' is a syzygy, i.e. since $\sum_{i=1}^{s} c_i b_i = 0$. ∎

It is easy to see that all special syzygies $\xi_l \epsilon_j$ appear only once in (5.11), while others can overlap.

Definition 5.3.5. Let $c_i \in \mathbb{k}^*, X_i \in \text{Mon}(\mathcal{A}), 1 \leqslant i \leqslant s$. Then we call a syzygy
$$\sum h_i \epsilon_i \in \text{Syz}(c_1 X_1, \ldots, c_s X_s)$$
homogeneous with exponent E if each h_i is a term satisfying
$$\text{Exp}(\text{Lm}(h_i)) \oplus \text{Exp}(X_i) = E$$
for all i such that $h_i \neq 0$.

Note that all elementary syzygy generators in Proposition 5.3.4 are homogeneous.

In order to to prepare for Theorem 5.3.6 below, let us recall some notations from Section 1.4.

Let $F = \{f_1, \ldots, f_s\} \subset \mathcal{R}^r \setminus \{0\}$, denote $\underline{F} := (f_1, \ldots f_s)$, $M := {}_\mathcal{R}\langle f_1, \ldots f_s \rangle$. Consider the module homomorphism
$$\psi_{\underline{F}} : \mathbf{F}_1 := \bigoplus_{i=1}^{s} \mathcal{R}\epsilon_i \twoheadrightarrow M \subset \mathcal{R}^r =: \mathbf{F}_0, \tag{5.16}$$
given by $\epsilon_i \mapsto f_i$.

5.3. SCHREYER ORDERING AND SYZYGIES OF LEADING TERMS

We are interested in a suitable generators for the syzygy module:

$$\mathrm{Syz}(\underline{F}) = \mathrm{Ker}(\psi_{\underline{F}}).$$

Let H be a homogeneous generating set of $\mathrm{Syz}(\mathrm{Lt}(f_1), \ldots, \mathrm{Lt}(f_s)) \subset \mathcal{A}^s \subset \mathcal{R}^s$. Consider for any syzygy of leading terms $S' = \sum g_i \epsilon_i \in H$, and its image under $\psi_{\underline{F}}$: $\psi_{\underline{F}}(S') = S' * \underline{F} = \sum g_i \cdot f_i \in M$.

Assume to have a weak normal form on $\mathcal{R}^r$ w.r.t. $\underline{F}$. If $\mathrm{NF}(S' * \underline{F} \mid F_{S'}) = 0$ for some $F_{S'} \subseteq F$ then there exists a standard representation

$$u \cdot (S' * \underline{F}) = \sum_{f_j \in F_{S'}} a_j \cdot f_j, \ a_j \in \mathcal{R}, \tag{5.17}$$

for some commutative unit $u \in \mathcal{R}$. Define

$$S := u \cdot S' - \sum_{j=1}^{s} a_j \epsilon_j.$$

Then $S \in \mathrm{Syz}(\underline{F})$ and $\mathrm{Lm}(S) = \mathrm{Lm}(uS') = \mathrm{Lm}(S')$ since $S' \in H \subset \mathrm{Syz}(\mathrm{Lt}(f_1), \ldots, \mathrm{Lt}(f_s))$ and due to properties of standard representation (5.17).

Denote the set of syzygies S arising as above from elements $S' \in H$ by $\widehat{H} \subseteq \mathrm{Syz}(\underline{F})$. Note that $\mathrm{L}(\widehat{H}) = \mathrm{L}(H)$.

Furthermore we assume that we have a weak normal form on $\mathcal{R}^s$ w.r.t. $\widehat{H}$.

The following theorem is one of the main results of this thesis. This theorem may be considered as a generalization of the Buchberger's Criterion to the case of a central localization of a graded commutative algebra.

Theorem 5.3.6. *Let $\mathcal{A}$ be a graded commutative algebra endowed with an ordering $>$ and $\mathcal{R} = \mathcal{A}_>$. Let $F = \{f_1, \ldots, f_s\} \subset \mathcal{R}^r \setminus \{0\}$, M be the left $\mathcal{R}$-module generated by $f_1, \ldots f_s$ and let $H \subset \mathcal{A}^s$ be a homogeneous generating set of $\mathrm{Syz}(\mathrm{Lt}(f_1), \ldots, \mathrm{Lt}(f_s))$ such that:*

1. $\forall S' \in H : \mathrm{NF}(S' * \underline{F} \mid F_{S'}) = 0$ *for some* $F_{S'} \subseteq F$.

Then the following statements hold:

1. $\mathrm{L}(M) \subseteq \mathrm{L}(F)$, *which follows that F is a SB of M w.r.t. $>$.*
2. $\mathrm{L}(\mathrm{Syz}(\underline{F})) \subseteq \mathrm{L}(\widehat{H})$, *which follows that $\widehat{H}$ is a SB of $\mathrm{Syz}(\underline{F})$ w.r.t. any Schreyer induced ordering. In particular, $\widehat{H}$ generates $\mathrm{Syz}(\underline{F})$.*

Proof. Take any $f \in M$ and a preimage $g \in \mathcal{R}^s$ of f:

$$g = \sum_{i=1}^{s} a_i \epsilon_i, f = \sum_{i=1}^{s} a_i f_i,$$

This is possible since F generates M.

We give a proof of 1 and 2 at the same time. For the former claim we assume $f \neq 0$, while for the later $f = 0$.

Consider a standard representation w.r.t. $\widehat{H}$:

$$ug = h + \sum_{S \in \widehat{H}} a_S S, a_S \in \mathcal{R}, \tag{5.18}$$

where u is a unit and h is a normal form of g w.r.t. $\widehat{H}$.

Note that if we apply ψ_F to (5.18) we will get

$$uf = \psi_F(h) = h * F, \tag{5.19}$$

If $h = 0$ then Claim 2 is clearly true due to properties of weak normal form but in this case: (5.19) follows that $f = 0$, which contradicts with the assumption $f \neq 0$ for Claim 1 and thus cannot happen.

Thus we may assume that $h = \sum_{j=1}^{s} h_j \epsilon_j \neq 0$ and let $\mathrm{Lm}(h) = \mathrm{Lm}(h_\nu) \epsilon_\nu$ for some $1 \leqslant \nu \leqslant s$.

Due to the properties of weak normal form we get that:

$$\mathrm{Lm}(h) \notin \mathrm{L}\left(\widehat{H}\right). \tag{5.20}$$

Recall that $\mathrm{L}\left(\widehat{H}\right) = \mathrm{L}(H)$ and $\xi_l \epsilon_\nu \in \mathrm{Syz}(\mathrm{Lt}(f_1), \ldots, \mathrm{Lt}(f_s))$ for all $\xi_l \mid \mathrm{Lm}(f_\nu)$. Thus $\mathrm{Lm}(h_\nu) \mathrm{Lm}(f_\nu) \neq 0$ since ξ_l dividing $\mathrm{Lm}(h_\nu)$ would contradict (5.20).

Due to (5.19) it follows that

$$uf = \sum_{j=1}^{s} h_j f_j. \tag{5.21}$$

We are going to show that

$$\mathrm{Lm}(f) = \pm \mathrm{Lm}(h_\nu) \mathrm{Lm}(f_\nu) \neq 0 \tag{5.22}$$

since this would show Claim 1 and provide a contradiction with the assumption $f = 0$ for Claim 2, which means that h must be zero.

In order to show (5.22) we need to consider cancelation between possible leading terms $\mathrm{Lm}(h_i f_i)$ with the same maximal exponent.

5.3. SCHREYER ORDERING AND SYZYGIES OF LEADING TERMS

Note that by definition of Schreyer induced ordering on $\mathcal{A}^s$ it follows that: $\operatorname{Exp}(\operatorname{Lm}(h_\nu)) \oplus \operatorname{Exp}(\operatorname{Lm}(f_\nu)) \geqslant \operatorname{Exp}(\operatorname{Lm}(h_i)) \oplus \operatorname{Exp}(\operatorname{Lm}(f_i))$, for all $1 \leqslant i \leqslant s$ such that $h_i \neq 0$. Thus no need to consider lower terms from $h_j f_j \neq 0$ with $\operatorname{Lm}(h_j) \operatorname{Lm}(f_j) = 0$ since in such a case $\operatorname{Exp}(\operatorname{Lm}(h_j f_j)) < \operatorname{Exp}(\operatorname{Lm}(h_j)) \oplus \operatorname{Exp}(\operatorname{Lm}(f_j)) \leqslant \operatorname{Exp}(\operatorname{Lm}(h_\nu)) \oplus \operatorname{Exp}(\operatorname{Lm}(f_\nu))$.

Assume that for some $j \neq \nu$ holds:

$$\pm \operatorname{Lm}(h_j) \operatorname{Lm}(f_j) = \pm \operatorname{Lm}(h_\nu) \operatorname{Lm}(f_\nu) =: M \in \operatorname{Mon}(\mathcal{A}^r).$$

Observe that $\operatorname{Lm}(f_\nu)$ and $\operatorname{Lm}(f_j)$ divide M and thus so also does

$$m := \operatorname{lcm}(\operatorname{Lm}(f_\nu), \operatorname{Lm}(f_j)) \in \operatorname{Mon}(\mathcal{A}^r).$$

Note that all module monomials $\operatorname{Lm}(f_\nu), \operatorname{Lm}(f_j), M$ and m involve the same module component in $\mathcal{A}^r$. Since m divides M it follows that $M/m \in \operatorname{Mon}(\mathcal{A})$. Moreover it follows that the monomial $m/\operatorname{Lm}(f_\nu) \in \operatorname{Mon}(\mathcal{A})$ divides $\operatorname{Lm}(h_\nu)$.

Consider the homogeneous elementary syzygy between $\operatorname{Lt}(f_\nu)$ and $\operatorname{Lt}(f_j)$:

$$S' = \operatorname{Lc}(f_j)\,(m /\!/ \operatorname{Lm}(f_\nu))\,\epsilon_\nu - \operatorname{Lc}(f_\nu)\,(m /\!/ \operatorname{Lm}(f_j))\,\epsilon_j \in \operatorname{Syz}(\operatorname{Lt}(f_1), \ldots, \operatorname{Lt}(f_s)).$$

Note that

$$\operatorname{Lm}((m /\!/ \operatorname{Lm}(f_j))\,\epsilon_j) = (m/\operatorname{Lm}(f_j))\,\epsilon_j, \quad \operatorname{Lm}((m /\!/ \operatorname{Lm}(f_\nu))\,\epsilon_\nu) = (m/\operatorname{Lm}(f_\nu))\,\epsilon_\nu.$$

If we assume that $\operatorname{Lm}(S') = (m/\operatorname{Lm}(f_j))\,\epsilon_j > (m/\operatorname{Lm}(f_\nu))\,\epsilon_\nu$ then by multiplying this inequality of module monomials with the monom M/m, we would get $\operatorname{Lm}(h_j)\epsilon_j > \operatorname{Lm}(h_\nu)\epsilon_\nu$ (no zero-division happens here!), which contradicts with above.

Therefore $\operatorname{Lm}(S') = (m/\operatorname{Lm}(f_\nu))\,\epsilon_\nu \in L(H)$ and it divides $\operatorname{Lm}(h)$, since $m/\operatorname{Lm}(f_\nu) \in \operatorname{Mon}(\mathcal{A})$ divides $\operatorname{Lm}(h_\nu)$. Hence we arrive to a contradiction with (5.20). ∎

Clearly our previous characterization of a SB (Theorem 3.4.6) is just an easy corollary from theorem 5.3.6.

Theorem 5.3.6 shows that in order to compute syzygies of original elements one only needs to consider syzygies of leading terms, which due to Proposition 5.3.4, are generated by elementary syzygies.

Moreover, one need only those elementary syzygies whose leading terms minimally generate the $L(\operatorname{Syz}(\operatorname{Lt}(f_1), \ldots, \operatorname{Lt}(f_s)))$ since they would give rise to a SB of $\operatorname{Syz}(f_1, \ldots, f_s)$ and thus to a GB of F. Thus we get the following lemma (5.3.7) which may be considered as a *generalized Chain Criterion*.

Lemma 5.3.7. *In the setting of Theorem 5.3.6 let $S_1, S_2 \in H$ such that $\operatorname{Lm}(S_1)$ divides $\operatorname{Lm}(S_2)$, $H' := H \setminus \{S_2\}$ and $\langle H' \rangle = \langle H \rangle$. Then $\operatorname{Lm}(S_2) \in L(H')$ and thus $\widehat{H'}$ is already a SB of Syzygies.*

Let us consider pairs of indices $1 \leqslant i < j \leqslant s$ such that $\mathrm{Lm}(f_i)$ and $\mathrm{Lm}(f_j)$ involve the same component. Denote $M_{ij} := \mathrm{lcm}(\mathrm{Lm}(f_i), \mathrm{Lm}(f_j)) \neq 0$ and $m_{ij} := (M_{ij} /\!/ \mathrm{Lm}(f_j))$. Let S_{ij} be the syzygy between $\mathrm{Lm}(f_i)$ and $\mathrm{Lm}(f_j)$:

$$S_{ij} = \mathrm{Lc}(f_j) m_{ij} \boldsymbol{\epsilon}_j - \mathrm{Lc}(f_j) m_{ji} \boldsymbol{\epsilon}_i.$$

With respect to the classical Schreyer ordering (cf. 5.3.3) it is clear that $\mathrm{Lm}(S_{ij}) = \mathrm{Lm}(m_{ij}) \boldsymbol{\epsilon}_j$. Note that the syzygy S_{ij} corresponds to our definition of $\mathrm{LeftSPoly}(f_i, f_j)$, i.e.

$$\mathrm{LeftSPoly}(f_i, f_j) = S_{ij} * \underline{F}$$

Therefore, we may say that w.r.t. the classical Schreyer ordering classical statements about classical elementary syzygies are true: for instance, our Lemma 5.3.7 implies an analog to lemma 2.5.10 (Chain Criterion) from [67]. Furthermore incorporating the resulting Chain Criterion, one may arrive to the following algorithm for computing Syzygies of leading terms, which is due to [81] and [32]:

Lemma 5.3.8. *Let us denote monomial ideals*

$$M_1 := \mathrm{Ann}(\mathrm{Lm}(f_1)) = \langle 0 \rangle : \mathrm{Lm}(f_1), \quad M_i := \langle \mathrm{Lm}(f_1), \ldots, \mathrm{Lm}(f_{i-1}) \rangle : \mathrm{Lm}(f_i), 1 < i \leqslant s.$$

Then with respect to the classical Schreyer ordering on $\mathrm{Mon}(\mathcal{A}^s)$ induced by $\mathrm{Lm}(f_1), \ldots, \mathrm{Lm}(f_s)$, the leading ideal of syzygies of leading terms, $\mathrm{L}(\mathrm{Syz}(\mathrm{Lt}(f_1), \ldots, \mathrm{Lt}(f_s)))$, is minimally generated by

$$\bigcup_{i=1}^{s} \mathrm{Mingens}(M_i) \boldsymbol{\epsilon}_i, \tag{5.23}$$

where $\mathrm{Mingens}(M_i)$ denotes the minimal set of generators of the monomial ideal M_i.

Remark 5.3.9. Moreover in the above Lemma 5.3.8 the following hold:

- $\mathrm{Ann}(\mathrm{Lm}(f_i)) = \langle \xi_l \mid \xi_l \cdot \mathrm{Lm}(f_i) = 0 \rangle \subseteq M_i$ for all i.
- the monomial ideal M_i is generated by the generators of $\mathrm{Ann}(\mathrm{Lm}(f_i))$ together with the following monomials: $\mathrm{lcm}(\mathrm{Lm}(f_k), \mathrm{Lm}(f_i))/\mathrm{Lm}(f_i)$ for all $1 \leqslant k < i$ such that $\mathrm{lcm}(\mathrm{Lm}(f_k), \mathrm{Lm}(f_i)) \neq 0$.

Remark 5.3.10. Note that in contrary to zero–division–free situation we need to start with $M_1 := \mathrm{Mingens}(0 : \mathrm{Lt}(f_1))$ since in our case $M_1 = \{\xi_l \mid \xi_l \mathrm{Lt}(f_1) = 0\}$.

Note that we incorporate Chain Criterion in lemma 5.3.8 since we only consider minimal generators of monomial ideals M_i.

The above Lemma 5.3.8 and Remark 5.3.9 heavily rely on the definition of classical Schreyer ordering and cannot be used without adjustments in the case of a general induced Schreyer ordering as the following examples illustrates.

Example 5.3.11. Let $A = \mathbb{k}[\xi, y, z]/\langle \xi^2 \rangle$ endowed with any monomial ordering $>_0$. Consider the pair of monomials: $X_1 := \xi y, X_2 := \xi z$.

Then the syzygy module $\mathrm{Syz}(X_1, X_2)$ is generated by $H := \{\xi \epsilon_1, \xi \epsilon_2, z \epsilon_1 - y \epsilon_2\}$ which is also a GB w.r.t. any Schreyer induced ordering $>_1$ on $\mathrm{Mon}(A^2)$. Thus minimal generators of $\mathrm{L}(H)$ are $\xi \epsilon_1, \xi \epsilon_2$ and $\mathrm{Lm}(z \epsilon_1 - y \epsilon_2)$.

If $>_1$ is the classical Schreyer ordering then $\mathrm{Lm}(z \epsilon_1 - y \epsilon_2) = y \epsilon_2$, since $2 > 1$, whereas for the classical reversed Schreyer ordering we get $\mathrm{Lm}(z \epsilon_1 - y \epsilon_2) = z \epsilon_1$.

This example shows that Lemma 3.5 from [81] assumes the classical Schreyer induced ordering and that $0 : t_1$ is zero.

Example 5.3.11 shows that for a general induced Schreyer ordering we cannot do better than actually consider all the elementary syzygies and choose some of them, whose leading terms minimally generate $\mathrm{L}(H)$. Notice that those chosen syzygies, is in fact everything what we need to know.

Lemma 5.3.12. *Let $f_1, \ldots, f_s \in \mathcal{A}^r$. Let us denote:*

$$
\begin{aligned}
m_{ij} &:= \mathrm{lcm}(\mathrm{Lm}(f_i), \mathrm{Lm}(f_j)) \in \mathrm{Mon}(\mathcal{A}^r) \cup \{0\} \\
S_{ij} &:= \mathrm{Lc}(f_j)\,(m_{ij}/\!\!/\mathrm{Lm}(f_i))\,\epsilon_i - \mathrm{Lc}(f_i)\,(m_{ij}/\!\!/\mathrm{Lm}(f_j))\,\epsilon_j, \\
\Sigma &:= \mathrm{Syz}(\mathrm{Lt}(f_1), \ldots, \mathrm{Lt}(f_s)) \subset \mathcal{A}^s, \\
H' &:= \mathrm{Interred}(\{S_{ij} \mid 1 \leqslant i < j \leqslant s\}) \subset \Sigma \\
H'' &:= \{\xi_l \epsilon_i \mid \xi_l \cdot \mathrm{Lm}(f_i) = 0, 1 \leqslant i \leqslant s\} \subset \Sigma \\
H &:= H' \cup H''
\end{aligned}
$$

where $\mathrm{Interred}(-)$ *is the usual interreduction of elementary (binomial) syzygies S_{ij} between each other w.r.t. $>_1$.*

Then with respect to any Schreyer ordering $>_1$ on $\mathrm{Mon}(\mathcal{A}^s)$ induced by $\mathrm{Lm}(f_1), \ldots, \mathrm{Lm}(f_s)$, the module of syzygies Σ is minimally generated by H and the monomial ideal $\mathrm{L}(\Sigma)$ is minimally generated by $\mathrm{Lm}(H)$.

Proof. Recall that elementary syzygies from H'' are distinct and do not overlap with the usual syzygies S_{ij}. The later might be redundant, exactly as in the commutative case, which calls for the interreduction procedure.

Note that the interreduction H' of $S' = \{S_{ij} \mid 1 \leqslant i < j \leqslant s\}$ has the following properties: H' generates $\langle S' \rangle$, i.e. $H' \subset \Sigma$, it is interreduced and $\mathrm{L}(H') = \mathrm{L}(S')$. It follows that H' minimally generates $\langle S' \rangle$. ∎

5.4 Computation of a free resolution

Recall that $\mathcal{A}$ is a graded commutative algebra (with odd variables $\xi_1, \ldots, \xi_m$) endowed with an ordering $>$ and $\mathcal{R} = \mathcal{A}_>$. It seems that everything holds for GR-algebras (or even for algebras over rings), provided one knows how to compute syzygies of leading terms (Algorithm 5.4.1).

Consider a sequence of finitely generated free left $\mathcal{R}$-modules $\mathbf{F}_i$ (no maps are given), which we like to denote by

$$\mathbf{F}_\bullet : \cdots \rightsquigarrow \mathbf{F}_{l+1} \rightsquigarrow \mathbf{F}_l \rightsquigarrow \cdots \rightsquigarrow \mathbf{F}_1 \rightsquigarrow \mathbf{F}_0. \tag{5.24}$$

We assume that each $\mathbf{F}_i$ is endowed with a fixed basis. Moreover we will say that $f \in \mathbf{F}_l$ has the level (or is of) level l, and denote level$(f) := l$.

In order to simplify algorithms let us introduce the notion of an elementary arrow associated to the sequence of free modules (5.24):

Definition 5.4.1. Let $l \in \mathbb{N}$. We call the tuple consisting of $g \in \mathbf{F}_{l+1}$ and $f \in \mathbf{F}_l$ an *elementary arrow* of level l and denote it by

$$\rho := (g \rightsquigarrow f).$$

Furthermore, for an elementary arrow $\rho = (g \rightsquigarrow f)$, we denote

$$\text{level}(\rho) := l, \text{target}(\rho) := f \in \mathbf{F}_l, \text{source}(\rho) := g \in \mathbf{F}_{l+1}.$$

Moreover, for two elementary arrows $(g \rightsquigarrow f), (g' \rightsquigarrow f')$ of the same level, we define:

$$a \cdot (g \rightsquigarrow f) + b \cdot (g' \rightsquigarrow f') := ((a \cdot g + b \cdot g') \rightsquigarrow (a \cdot f + b \cdot f')),$$

for any $a, b \in \mathcal{R}$.

Definition 5.4.2. Let us denote the direct sum of all free modules occurring in $\mathbf{F}_\bullet$ by:

$$\mathbf{F}_\star := \bigoplus_{l \geq 0} \mathbf{F}_l,$$

which is clearly graded by l. In what follows we only consider homogeneous $f \in \mathbf{F}_\star$, that is, $\exists l : f \in \mathbf{F}_l$ and thus we will omit the word "homogeneous" and extend the notion of level to these elements by putting level$(f) := l$.

The union of the bases of the $\mathbf{F}_l$ is a basis of $\mathbf{F}_\star$, which we denote for the future reference by $\{\epsilon_j, j \in \mathbb{N}\}$.

We denote the left $\mathcal{R}$-module of all elementary arrows of level l by $\text{Hom}_\mathcal{R}(\mathbf{F}_{l+1} \to \mathbf{F}_l)$. The direct sum of all such modules shall be denoted by

$$\text{Hom}_\mathcal{R}(\mathbf{F}_\star \to \mathbf{F}_\star) := \bigoplus_{l \geq 0} \text{Hom}_\mathcal{R}(\mathbf{F}_{l+1} \to \mathbf{F}_l),$$

which is graded by l. In what follows we only consider homogeneous $\rho \in \text{Hom}_\mathcal{R}(\mathbf{F}_\star \to \mathbf{F}_\star)$, that is, $\exists l : \rho \in \text{Hom}_\mathcal{R}(\mathbf{F}_{l+1} \to \mathbf{F}_l)$ and thus we will omit the word "homogeneous" and extend the notion of level to these elements by putting level$(\rho) := l$.

5.4. COMPUTATION OF A FREE RESOLUTION

Let $\Phi \subset \mathrm{Hom}_{\mathcal{R}}(\mathbf{F}_\star \to \mathbf{F}_\star)$ be a finite set of elementary arrows. We extend the notions of target and source to sets of elementary arrows by putting:

$$\mathrm{target}(\Phi) := \{\mathrm{target}(\rho) \mid \rho \in \Phi\} \subset \mathbf{F}_\star, \quad \mathrm{source}(\Phi) := \{\mathrm{source}(\rho) \mid \rho \in \Phi\} \subset \mathbf{F}_\star.$$

The left submodule generated by Φ in $\mathrm{Hom}_{\mathcal{R}}(\mathbf{F}_\star \to \mathbf{F}_\star)$ shall be denoted as follows:

$$_{\mathcal{R}}\langle \Phi \rangle := \left\{ \sum_{\rho \in \Phi} a_\rho \cdot \rho, a_\rho \in \mathcal{R} \right\} \subset \mathrm{Hom}_{\mathcal{R}}(\mathbf{F}_\star \to \mathbf{F}_\star).$$

Let us now recall that $\mathcal{R}$ is the central localization of a graded commutative algebra $\mathcal{A}$, to whose odd elements we will refer to as $\xi_1, \ldots, \xi_m$. Moreover, we assume that each free $\mathcal{R}$-module $\mathbf{F}_i$ is endowed with a monomial ordering compatible with it on $\mathcal{A}$.

Definition 5.4.3. Let $\rho = (g \rightsquigarrow f) \in \mathrm{Hom}_{\mathcal{R}}(\mathbf{F}_{l+1} \to \mathbf{F}_l), \Phi \subset \mathrm{Hom}_{\mathcal{R}}(\mathbf{F}_\star \to \mathbf{F}_\star)$. Denote:

$$\mathrm{Lm}(\rho) := \mathrm{Lm}(f) \in \mathrm{Mon}(\mathbf{F}_l), \quad \mathrm{Lt}(\rho) := \mathrm{Lt}(f) \in \mathbf{F}_l, \quad \mathrm{Lc}(\rho) := \mathrm{Lc}(f) \in \Bbbk,$$
$$\mathrm{Lm}(\Phi) := \{(g \rightsquigarrow \mathrm{Lm}(f)) \mid (g \rightsquigarrow f) \in \Phi\} \subset \mathrm{Hom}_{\mathcal{R}}(\mathbf{F}_\star \to \mathbf{F}_\star),$$
$$\mathrm{Lt}(\Phi) := \{(g \rightsquigarrow \mathrm{Lt}(f)) \mid (g \rightsquigarrow f) \in \Phi\} \subset \mathrm{Hom}_{\mathcal{R}}(\mathbf{F}_\star \to \mathbf{F}_\star).$$

Let $\Phi, \Lambda \subset \mathrm{Hom}_{\mathcal{R}}(\mathbf{F}_{i+1} \to \mathbf{F}_i)$. We call Φ a *lifting* of Λ if $\mathrm{Lm}(\Phi) = \Lambda$.

Definition 5.4.4. We extend the notion of (left) S-polynomials to elementary arrows. Let $\rho = (g \rightsquigarrow f)$ and $\rho' = (g' \rightsquigarrow f')$ be two elementary arrows of the same level and $a, b \in \mathcal{R}$ such that $\mathrm{LeftSPoly}(f, f') = a \cdot f + b \cdot f'$, then

$$\mathrm{LeftSPoly}(\rho, \rho') := ((a \cdot g + b \cdot g') \rightsquigarrow \mathrm{LeftSPoly}(f, f')).$$

Definition 5.4.5. Let $\Phi = \{(\epsilon_{i_j} \rightsquigarrow f_j) \mid i_j \in \mathbb{N}\}$ be a finite subset of $\mathrm{Hom}_{\mathcal{R}}(\mathbf{F}_\star \to \mathbf{F}_\star)$. It can be extended to a homomorphism from a free submodule $\bigoplus_j \mathcal{R}\epsilon_{i_j} \subset \mathbf{F}_\star$ into $\mathbf{F}_\star$:

$$\psi_\Phi : \bigoplus_j \mathcal{R}\epsilon_{i_j} \to \mathbf{F}_\star : \epsilon_j \mapsto f_j. \tag{5.25}$$

We denote the kernel of ψ_Φ as follows:

$$\mathrm{Syz}(\Phi) := \mathrm{Ker}(\psi_\Phi) \subset \mathbf{F}_\star.$$

Due to Lemma 5.3.12 we can compute "leading syzygies" via Algorithm 5.4.1.

Algorithm 5.4.1 LEADING_SYZYGIES(Λ)

ASSUME: odd variables of $\mathcal{R}$ are denoted by $\xi_1, \ldots, \xi_m$
INPUT: Λ a finite subset of $\mathrm{Hom}_{\mathcal{R}}(\mathbf{F}_\star \to \mathbf{F}_\star)$: $\mathrm{target}(\Lambda)$ consists of terms
OUTPUT: a finite generating set of $\mathrm{Syz}(\Lambda)$
1: $H' = \{\mathrm{source}(\mathrm{LeftSPoly}(\tau, \tau')) \mid \tau, \tau' \in \Lambda, \mathrm{level}(\tau) = \mathrm{level}(\tau')\}$;
2: $H'' = \{\xi_k \cdot g \mid (g \rightsquigarrow f) \in \Lambda, \xi_k \cdot \mathrm{Lm}(f) = 0\}$;
RETURN: $H' \cup H''$;

Let Φ be as in Definition 5.4.5. Consider $\Lambda = \mathrm{Lt}(\Phi)$ and $H = \mathrm{LEADING\_SYZYGIES}(\Lambda) \subset \mathbf{F}_\star$ via Algorithm 5.4.1. Then for any $s = \sum_j a_j \boldsymbol{\epsilon}_{i_j} \in H$ we can compute its image under ψ_Φ, and denote it by:

$$s * \Phi := \psi_\Phi(s) = \sum_j a_j \cdot f_j.$$

Furthermore, all our reduction algorithms can be extended to work with elementary arrows. For example: Algorithm 3.3.1 can be extended as in Algorithm 5.4.2

Algorithm 5.4.2 LEFTNF(ρ', Φ)

INPUT: $\rho' \in \mathrm{Hom}_\mathcal{R}(\mathbf{F}_{l+1} \to \mathbf{F}_l)$, Φ a finite subset of $\mathrm{Hom}_\mathcal{R}(\mathbf{F}_\star \to \mathbf{F}_\star)$;
OUTPUT: $\rho \in \mathrm{Hom}_\mathcal{R}(\mathbf{F}_{l+1} \to \mathbf{F}_l)$ such that $\mathrm{target}(\rho)$ is a left normal form of $\mathrm{target}(\rho')$ w.r.t. $\{\mathrm{target}(f) : f \in \Phi, \mathrm{level}(f) = l\}$ and $\mathrm{source}(\rho)$ is the corresponding *history of reductions*
1: $\rho := \rho'$;
2: **while** $(\mathrm{target}(\rho) \neq 0)$ and $(\Gamma_\rho := \{\tau \in \Phi : \mathrm{level}(\tau) = l, \mathrm{Lm}(\tau) | \mathrm{Lm}(\rho)\} \neq \emptyset)$ **do**
3:     Choose any $\tau \in \Gamma_\rho$;
4:     $\rho := \mathrm{LeftSPoly}(\rho, \tau)$;
5: **end while**
RETURN: ρ;

We extend the notion of ecart by putting: $\mathrm{Ecart}(\rho) := \mathrm{Ecart}(\mathrm{target}(\rho))$.

Algorithm LEFTNFMORA (as well as the interreduction procedure) can be extended analogously to our extension of LEFTNF (Algorithm 5.4.2).

Using our notations in the context of the sequence (5.16) we propose a variation of the usual Buchberger's algorithm which computes a SB and, as a byproduct, a syzygy module of input: Algorithm 5.4.3.

5.4. COMPUTATION OF A FREE RESOLUTION

Algorithm 5.4.3 BBA_SYZ(Φ)
ASSUME: $NF(-\mid-)$ is any normal form (e.g. LEFTNF or LEFTNFMORA)
INPUT: $\Phi = \{(\epsilon_i \rightsquigarrow f_i) \mid 1 \leqslant i \leqslant s\} \subset \mathrm{Hom}_{\mathcal{R}}(\mathbf{F}_1 \to \mathbf{F}_0)$;
OUTPUT: (G, S), where G a SB of ${}_{\mathcal{R}}\langle f_i \rangle$ and $S \subset \mathbf{F}_1$ generates $\mathrm{Syz}(\Phi)$
1: $S := \emptyset, \Gamma := \Phi$;
2: $H := \mathrm{Interred}(\mathrm{LEADING\_SYZYGY}(\mathrm{Lt}(\Gamma))) \subset \mathbf{F}_1$; // via Algorithm 5.4.1
3: **while** $\exists s \in H$ **do**
4: $\quad H := H \setminus \{s\}$;
5: $\quad \rho := (s \rightsquigarrow s * \Phi)$; // general S-polynomial in terms of Φ
6: $\quad (g \rightsquigarrow f) := \mathrm{NF}(\rho \mid \Gamma)$;
7: $\quad$ **if** $f \neq 0$ **then**
8: $\quad\quad$ add to H new syzygies involving new leading term $\mathrm{Lt}(f)$; // As in Algorithm 5.4.5
9: $\quad\quad \Gamma := \Gamma \cup \{(g \rightsquigarrow f)\}$;
10: $\quad$ **else**
11: $\quad\quad S := S \cup \{g\}$;
12: $\quad$ **end if**
13: **end while**
RETURN: $(\mathrm{target}(\Gamma), S)$;

The termination of Algorithm 5.4.3 follows as usually from the impossibility of infinite growth of $L(G)$, which is bounded by $L({}_{\mathcal{R}}\langle\mathrm{target}(\Phi)\rangle)$ and finitely generated by Dickson's Lemma. The correctness follows from Theorem 5.3.6.

Now we start to prepare for the methods analogous to Schreyer's and La Scala's.

For the sake of simplicity let

$$F_0 = \{f_1, \ldots, f_{r_1}\} \subset \mathbf{F}_0 := \bigoplus_{i=1}^{r_0} \mathcal{R}\epsilon_i$$

be an interreduced SB of

$$M := {}_{\mathcal{R}}\langle F_0 \rangle$$

w.r.t. a given module ordering $>_0$ on $\mathbf{F}_0$. We set

$$\mathbf{F}_1 := \bigoplus_{i=1}^{r_1} \mathcal{R}\epsilon_{r_0+i},$$
$$\Phi_0 := \{(\epsilon_{r_0+i} \rightsquigarrow f_i), 1 \leqslant i \leqslant r_1\} \subset \mathrm{Hom}_{\mathcal{R}}(\mathbf{F}_1 \to \mathbf{F}_0),$$
$$\Lambda_0 := \mathrm{Lm}(\Phi_0) \subset \mathrm{Hom}_{\mathcal{R}}(\mathbf{F}_1 \to \mathbf{F}_0).$$

Let $>_1$ be any Schreyer ordering on $\mathrm{Mon}(\mathbf{F}_1) = \mathrm{Mon}(\mathcal{A}^{r_1})$ induced by Λ_0, that is, in fact by $\underline{L_0} := (\mathrm{Lm}(f_1), \ldots, \mathrm{Lm}(f_{r_1})) \in (\mathbf{F}_0)^{r_1}$.

Next we compute a minimal generating set of $L(\mathrm{Syz}(\underline{L_0}))$ in terms of $\epsilon_{r_0+1}, \ldots, \epsilon_{r_0+r_1}$ and order it w.r.t. $>_1$. The result shall be denoted by $\underline{L_1}$. We abbreviate this procedure by writing
$$\underline{L_1} := \mathrm{SortedMinbase}(L(\mathrm{Syz}(\Lambda_0))) = (b_1, \ldots, b_{r_2}) \in (\mathbf{F}_1)^{r_2}.$$

We set
$$\mathbf{F}_2 := \bigoplus_{i=1}^{r_2} \mathcal{R}\epsilon_{r_0+r_1+i}, \quad \Lambda_1 := \{(\epsilon_{r_0+r_1+i} \rightsquigarrow b_i), 1 \leqslant i \leqslant r_2\} \in \mathrm{Hom}_\mathcal{R}(\mathbf{F}_2 \to \mathbf{F}_1).$$

Therefore, $L_1 = \{b_1, \ldots, b_{r_2}\}$ is a minimal generating set of $L(\mathrm{Syz}(\Phi_0))$ (due to Theorem 5.3.6 and Lemma 5.3.12).

Iteratively repeating the above construction further we obtain the sequence (no maps):
$$\mathbf{F}_\bullet : \cdots \rightsquigarrow \mathbf{F}_l = \bigoplus_{i=1}^{r_l} \mathcal{R}\epsilon_{s_{l-1}+i} \rightsquigarrow \mathbf{F}_{l-1} \rightsquigarrow \cdots \rightsquigarrow \mathbf{F}_1 \rightsquigarrow \mathbf{F}_0, \tag{5.26}$$

where $\mathbf{F}_i$ is a free $\mathcal{R}$-module of rank r_i, endowed with an induced Schreyer ordering $>_i$ on the corresponding set of module monomials $(0 < i)$ and $s_0 := 0, s_l := \sum_{j=0}^{l} r_j$ for $l > 0$. Moreover, due to our construction it follows that:

$$\mathbf{F}_\star = \bigoplus_{l \geqslant 0} \bigoplus_{i=1}^{r_l} \mathcal{R}\epsilon_{s_{l-1}+i}.$$

Furthermore, the following leading data correspondence have been computed in the process:
$$\Lambda_i \subset \mathrm{Hom}_\mathcal{R}(\mathbf{F}_i \to \mathbf{F}_{i-1}), \quad L_i \subset \mathrm{Mon}(\mathbf{F}_i).$$

Note that we can already fill in the first gap between modules: the map $\mathbf{F}_1 \xrightarrow{\psi_{\Phi_0}} \mathbf{F}_0$ (recall the map ψ_F from (5.16)), satisfies $\mathrm{Coker}(\psi_{\Phi_0}) = M$. Moreover, by construction $L(\mathrm{Ker}(\psi_{\Phi_0}))$ is minimally generated by L_1.

Since F_0 is a SB, the syzygies S computed by Algorithm 5.4.3 form a SB of $\mathrm{Syz}(\Phi_0)$ w.r.t. $>_1$ due to Theorem 5.3.6. Since $L(\mathrm{Syz}(\Phi_0)) = L_1$ by interreducing elements from S between each other we obtain an interreduced SB of $\mathrm{Syz}(\Phi_0)$, which we now denote by $F_1 \subset \mathbf{F}_1$. Note that we also reorder these elements: $\underline{F_1} = (f_{r_1+1}, \ldots, f_{r_1+r_2})$ accordingly to the order of monomials in $\underline{L_1}$, that is, where $\mathrm{Lm}(f_{r_1+i})$ is the i-th entry of $\underline{L_1}$.

Hence we have lifted Λ_1 to
$$\Phi_1 := \{(\epsilon_{r_0+r_1+i} \rightsquigarrow f_{r_1+i}), 1 \leqslant i \leqslant r_2\} \subset \mathrm{Hom}_\mathcal{R}(\mathbf{F}_2 \to \mathbf{F}_1),$$

which defines the homomorphisms $\mathbf{F}_2 \xrightarrow{\psi_{\Phi_1}} \mathbf{F}_1$ for the sequence (5.26), such that $\mathrm{target}(\Phi_1)$ is an interreduced SB of $\mathrm{Ker}(\psi_{\Phi_0})$.

5.4. COMPUTATION OF A FREE RESOLUTION

By repeating the above lifting procedure we can construct a free resolution of $\mathbf{F}_0/M$, which is known as **Schreyer resolution** (cf. [67]).

By analogy with $\mathbf{F}_i$ and $\mathbf{F}_\star$ we will always use the following notation for sets, graded by level:

$$\Lambda_\star := \bigoplus_{l \geq 0} \Lambda_l, \quad L_\star := \bigoplus_{l \geq 0} L_l, \quad \Phi_\star := \bigoplus_{l \geq 0} \Phi_l.$$

By construction it is clear that $\text{target}(\Lambda_\star) = L_\star$.

Definition 5.4.6. In the spirit of [81], we call $\Lambda_\star$ a **Schreyer frame** of $\mathbf{F}_0/M$, while its lifting $\Phi_\star$ (i.e. $\text{Lm}(\Phi_i) = \Lambda_i$) such that $\Phi_{i+1} = \text{Syz}(\Phi_i)$, $M = {}_\mathcal{R}\langle\text{target}(\Phi_0)\rangle$ shall be called a **Schreyer resolution** of $\mathbf{F}_0/M$, since the sequence $\mathbf{F}_\bullet$ together with the homomorphisms $\mathbf{F}_{i+1} \xrightarrow{\psi_{\Phi_i}} \mathbf{F}_i$, is indeed a free resolution of $\mathbf{F}_0/M$.

Note that we have just shown the classical (iterative) way to construct a Schreyer resolution (i.e. iterative lifting). Following [81] we propose to consider the process of filling up a Schreyer frame $\Lambda_\star$ to a Schreyer resolution $\Phi_\star$ by starting with Φ_0. This will be a generalization of Algorithm 5.4.3 in the sense that we use $\Phi_\star$ to compute elementary syzygies of leading terms $\Pi_\star$, which are than used to compute new elements of $\Phi_\star$ and so on. Due to the freedom of choice of next elementary syzygy this process in not necessarily iterative level by level. The Schreyer frame, fixed at the beginning, serves to "place" new elements correctly to their allotted "gaps" in the resulting Schreyer resolution.

We sketch our approach in Algorithm 5.4.4, which is a variation of Algorithm 4.1 RESOLUTION from [81]. The advantage of our variation is the consistent use of whole syzygies of leading terms (i.e. $\text{Syz}(\text{Lt}(\Phi_\star))$) together with the procedure for updating the current set of syzygies $\Pi_\star$ over using only their leading monomials (i.e. $\text{Lm}(\text{Syz}(\text{Lt}(\Phi_\star)))$) as in [81] ($H = \cup H_i$).

Algorithm 5.4.4 may be understood as a generalization of the original RESOLUTION Algorithm from [81] to work over central localizations of graded commutative algebras, which of course include commutative polynomial algebras. In order to tackle the case of mixed ordering we make use of polynomial weak normal forms introduced in Chapter 4.

Algorithm 5.4.4 LASCALA_RESOLUTION($F, \Lambda_\star$)

ASSUME: Let $>_0$ be an ordering on $\mathrm{Mon}(\mathbf{F}_0)$ such that $\xi_i > 1$, $\mathrm{NF}(\cdot \mid \cdot)$ a weak NF
INPUT: $F = (f_1, \ldots, f_s) \in \mathcal{G}_{\mathbf{F}_0}$ an interreduced SB of $M := {}_\mathcal{R}\langle f_1, \ldots, f_s\rangle$, $\Lambda_\star$ a Schreyer frame of $\mathbf{F}_0/M$;
OUTPUT: $\Phi_\star$ a Schreyer resolution of $\mathbf{F}_0/M$, $\Delta_\star \subset \Phi_\star$
1: $\Delta_i := \Phi_i := \Pi_i := \emptyset$ for all $i \geqslant 0$;
2: $\Delta_0 := \{(\epsilon_i \leadsto f_i) \mid 1 \leqslant i \leqslant s\} \subset \mathrm{Hom}_\mathcal{R}(\mathbf{F}_1 \to \mathbf{F}_0)$;
3: $\Phi_0 := \Delta_0$; // Note: $\mathrm{Lm}(\Phi_0) = \Lambda_0$
4: $H := \mathrm{LEADING\_SYZYGY}(\mathrm{Lt}(\Phi_0)) \subset \mathbf{F}_1$; // via Algorithm 5.4.1
5: $H := \mathrm{Interred}(H) \subset \mathbf{F}_1$; // Note: $\mathrm{L}(H) = L_1$
6: $\Pi_1 := \{(\epsilon \leadsto s) \mid s \in H, (\epsilon \leadsto \mathrm{Lm}(s)) \in \Lambda_1\}$; // Note: $\mathrm{Lm}(\Pi_1) = \Lambda_1$
7: **while** $\Pi_\star := \bigoplus_{l \geqslant 0} \Pi_l \neq \emptyset$ **do**
8:    Choose $(\epsilon \leadsto s) \in \Pi_\star$ with $\mathrm{Lm}(s)$ minimal;
9:    $i := \mathrm{level}(s)$; // Note: $\mathrm{Lm}(s) \in L_i$
10:    $\Pi_i := \Pi_i \setminus \{(\epsilon \leadsto s)\}$;
11:    $\rho := s * \Phi_{i-1}$; // general S-polynomial corresponding to syzygy of leading terms s
12:    $(g \leadsto f) := \mathrm{NF}(\rho \mid \Phi_{i-1})$; // Note: $\mathrm{Lm}(g) = \mathrm{Lm}(u \cdot s) = \mathrm{Lm}(s)$, $\mathrm{Lm}(u) = 1$
13:    **if** $f = 0$ **then**
14:      $\rho' := (\epsilon \leadsto g)$;
15:      $\Delta_i := \Delta_i \cup \{\rho'\}$;
16:    **else**
17:      Find $\epsilon'\colon (\epsilon' \leadsto \mathrm{Lm}(f)) \in \Lambda_{i-1}$; // Possible since $\mathrm{Lm}(f) \in L_{i-1}$
18:      Mark $(\epsilon' \leadsto \mathrm{Lm}(f))$; // it shall never occur in $\mathrm{Lm}(\Pi_{i-1})$
19:      $\tau := (\epsilon' \leadsto f)$;
20:      $\Pi_i := \Pi_i \cup \mathrm{NEW\_SYZYGIES}(\Pi_i, (\epsilon' \leadsto \mathrm{Lt}(f)), \mathrm{Lt}(\Phi_{i-1}))$; // $\mathrm{L}(\Pi_i) \subset \Lambda_i$
21:      $\Phi_{i-1} := \Phi_{i-1} \cup \{\tau\}$; // Note: $\mathrm{Lm}(\Phi_{i-1}) \subset \Lambda_{i-1}$
22:      $\rho' := (\epsilon \leadsto g - \epsilon')$; // Note: $\mathrm{Lm}(g) >_i \epsilon'$
23:    **end if**
24:    $\Pi_{i+1} := \Pi_{i+1} \cup \mathrm{NEW\_SYZYGIES}(\Pi_{i+1}, (\epsilon \leadsto \mathrm{Lt}(g)), \mathrm{Lt}(\Phi_i))$; // $\mathrm{L}(\Pi_{i+1}) \subset \Lambda_{i+1}$
25:    $\Phi_i := \Phi_i \cup \{\rho'\}$; // Note: $\mathrm{Lm}(\Phi_i) \subset \Lambda_i$, $\mathrm{Lm}(\mathrm{Syz}(\mathrm{Lt}(\Phi_i))) = \mathrm{Lm}(\Pi_{i+1}) \subset \Lambda_{i+1}$
26: **end while**
RETURN: $\Phi_\star, \Delta_\star$;

Algorithm 5.4.4 uses Algorithm 5.4.5 for updating Π_{i+1} in order to add new elementary syzygies involving a newly found element, which is about to be put into the corresponding "running SB" Φ_i.

5.4. COMPUTATION OF A FREE RESOLUTION

Algorithm 5.4.5 NEW_SYZYGIES(Π, ρ, Ξ)

ASSUME: This algorithm is called from 5.4.4 in the context of a Schreyer frame $\Lambda_\star$
INPUT: $\Pi \subset \operatorname{Syz}(\Xi)$ interreduced, $\rho = (\epsilon \leadsto X)$ of the same level as elements from Ξ
OUTPUT: $\Sigma \subset \operatorname{Syz}(\Xi \cup \{\rho\})$ interreduced and reduced w.r.t. Π, $\operatorname{Lm}(\Sigma) \subset \Lambda_\star$
1: $S := \{\operatorname{source}(\operatorname{LeftSPoly}(\tau, \rho)) \mid \tau \in \Xi\}$; // cf. Algorithm 5.4.1
2: $S := S \cup \{\xi_k \cdot \epsilon \mid \xi_k \cdot X = 0\}$; // all elementary syzygies involving ρ
3: $S := \operatorname{NF}(\operatorname{Interred}(S) \mid \operatorname{target}(\Pi))$; // Note: $\forall s \in S : \operatorname{Lm}(s) \in L_\star$
4: $\Sigma := \{(\epsilon' \leadsto s) \mid s \in S, (\epsilon' \leadsto \operatorname{Lm}(s)) \in \Lambda_\star - \text{not \underline{marked}}\}$; // Possible by above note
RETURN: Σ;

Remark 5.4.7. Showing correctness of Algorithm 5.4.4 is analogous to Proposition 4.4 from [81] and mainly follows from the line comments in Algorithms 5.4.4 and 5.4.5. The main point is that for new elements τ and ρ', $\operatorname{Lm}(\tau)$ and $\operatorname{Lm}(\rho')$ indeed belong to $L_\star$. This can be proved by observing that the new elements are reduced w.r.t. the corresponding $\Phi_\star$, that $\Lambda_\star$ is interreduced and the inclusion $\operatorname{Lm}(\Phi_\star) \subset \Lambda_\star$ still holds after enriching $\Phi_\star$ (induction on iterations).

Similary new elementary syzygies are interreduced and reduced w.r.t. to the old ones. Therefore $\operatorname{Lm}(\Sigma) \subset \Lambda_\star$.

Algorithm 5.4.5 only terminates if we impose some additional conditions, e.g. the length of the resolution, that is the maximal level of choose-able syzygies of leading terms. In the graded case one may be only interested in syzygies up to some maximal (induced) degree.

Note that by ordering the leading monomials from L_l ascending w.r.t. $>_l$ while constructing Schreyer resolution we may assume that the next induced ordering $>_{l+1}$ is in fact the classical Schreyer ordering. Moreover, while doing this we define the inductive sum of exponents $|\epsilon_i| \in \Upsilon$ for a module component ϵ_i of level $l+1$ corresponding to monomial $m\epsilon_j \in L_l$ inductively by putting $|\epsilon_i| := |m\epsilon_j| = \operatorname{Exp}(m) \oplus |\epsilon_j| \in \Upsilon$, with the starting initialization for the given canonical basis of $\mathbf{F}_0$: $|\epsilon_i| := \operatorname{Exp}(\epsilon_i) \in \Upsilon, 1 \leqslant i \leqslant r_0$ (here we recall Section 5.2).

That is in order to be able to compare two monomials $m_1\epsilon_{i_1}, m_2\epsilon_{i_2}$ of the same level $l+1$ we compare sums of exponents $\operatorname{Exp}(m_1) \oplus |\epsilon_{i_1}|$ and $\operatorname{Exp}(m_2) \oplus |\epsilon_{i_2}|$ w.r.t. the assumed in Section 5.2 total ordering on Υ. In the tie case the bigger module component (e.g. $i_1 > i_2$) wins.

The selection strategy is defined by the ordering of the elements from $L_\star = \bigoplus_{i \geqslant 0} L_i$, where each L_i is interreduced by construction. One such strategy was already used for computing a Schreyer resolution iteratively level by level: it gave priority to elementary syzygies between elements of least level.

The main advantages of such a freedom of choice is the possibility to apply the so called ***improvement after La Scala***: in the `else` branch of Algorithm 5.4.4 we get two new syzygies in one go. In order to use these improvements we shall make the selection strategy go to higher level as soon as computed elements from $\Phi_\star$ allow this.

Remark 5.4.8. Let us consider the graded case. That is, a degree function w-Deg$(-)$ defined on Υ is given and extended to $\mathrm{Mon}(\mathbf{F}_\star) \ni m\epsilon$ inductively by putting w-Deg$(m\epsilon) :=$ w-Deg$(|m\epsilon|)$ and the input elements are graded: $\forall f \in F_0 :$ w-Deg$(m\epsilon)$ is the same for all $m\epsilon \in \mathrm{Mon}(f)$. Note that all our operations on elements respect the grading. This means that all the syzygies, all the elements become (inductively) graded resulting in a graded free resolution.

Now, following [81], we consider a selection strategy (called ***minimization strategy***) satisfying the following condition: if for $n, m \in L_\star$ either

(w-Deg(m) $-$ level(m) $\leqslant$ w-Deg(n) $-$ level(n) and level$(m) <$ level(n))

or (w-Deg$(m) <$ w-Deg(n) and level$(m) =$ level(n))

or (w-Deg$(m) =$ w-Deg(n) and level$(m) >$ level(n))

then it must follow that $m < n$.

As in [81], it can be shown that a minimal Schreyer resolution of M can be computed with the use of a minimization strategy by using the "minimal" elements from $\Delta_\star$ and elimination the rest by substituting, say ϵ' with corresponding g (cf. Algorithm 5.4.4 the `else` branch).

Chapter 6

Graded commutative algebras in SINGULAR

This thesis is written with the aim of developing algorithms. We have extended and further developed the non-commutative subsystem of SINGULAR (SINGULAR:PLURAL) into a framework, which makes it possible to seamlessly embed algebra specific efficient algorithms, as we did for our central localizations of graded commutative algebras.

On the implementation side we have further developed the SINGULAR non-commutative subsystem SINGULAR:PLURAL in order to allow polynomial arithmetic and more involved non-commutative basic Computer Algebra computations (e.g. S-polynomial, GB) to be easily implementable for specific algebras. At the moment graded commutative algebra-related algorithms are implemented in this framework. The developed framework is briefly described in this chapter.

6.1 High level interface - users manual

Since a graded commutative algebra can be represented as a GR-algebra one defines it in SINGULAR just as one, that is, one has to construct a commutative ring, turn it into an anti-commutative algebra with the `nc_algebra` command and use it to form a quotient algebra with the `qring` command.

General SINGULAR User's Manual can be found online at [118]. SINGULAR's general non-commutative subsystem (including G- and GR-algebras) is also described in [86]. Here we present only the commands, which are specific to graded commutative algebras.

The detection of the graded commutative structure, which is done upon the `qring` command, automatically changes the internal general (GR-algebra) implementation to our special graded commutative one, requiring no additional user commands.

Note that monomial ordering is induce by it on the original commutative polynomial ring.

CHAPTER 6. GRADED COMMUTATIVE ALGEBRAS IN SINGULAR

Let us for example construct a graded commutative $\mathbb{Q}$-algebra A with 2 commutative variables, called x(1), x(2) and 2 anti-commuting variables, called y(1), y(2), endowed with an ordering induced by a degree-reverse-lexicographical ordering on $\mathrm{Mon}(x(1), y(1), y(2), x(2))$. The corresponding SINGULAR code looks as follows:

```
ring R = 0,(x(1), y(1..2), x(2)),(dp);
matrix C[4][4] = 1,1,1,1, 0,1,-1,1, 0,0,1,1, 0,0,0,1;
print(C);
↦ 1,1,1, 1,
↦ 0,1,-1,1,
↦ 0,0,1, 1,
↦ 0,0,0, 1
def S = nc_algebra(C,0); setring S;
S; // Needed anti-commutative G-algebra
↦ //   characteristic : 0
↦ //   number of vars : 4
↦ //        block   1 : ordering dp
↦ //                  : names    x(1) y(1) y(2) x(2)
↦ //        block   2 : ordering C
↦ //   noncommutative relations:
↦ //       y(2)y(1)=-y(1)*y(2)
y(2)*y(1); // x(i) are commutative and y(i) anti-commute,
↦ -y(1)*y(2)
y(1)*y(1); // Observe that S have no zero-divisors yet
↦ y(1)^2
ideal Q = y(1)^2, y(2)^2; // Squares
qring A = twostd(Q);
A; // this is a graded commutative algebra with needed variables
↦ //   characteristic : 0
↦ //   number of vars : 4
↦ //        block   1 : ordering dp
↦ //                  : names    x(1) y(1) y(2) x(2)
↦ //        block   2 : ordering C
↦ //   noncommutative relations:
↦ //       y(2)y(1)=-y(1)*y(2)
↦ // quotient ring from ideal
↦ _[1]=y(2)^2
↦ _[2]=y(1)^2
y(2)*y(1);
↦ -y(1)*y(2)
y(1)*y(1); // desired zero
↦ 0
```

Note that at the moment our SINGULAR implementation supports only one continuous block of "global" anti-commuting variables. For example, the graded commutative algebra A constructed in SINGULAR as follows – uses the general GR-algebra implementation:

```
ring R = 0,(y(1), z, y(2)),(dp);
matrix C[3][3] = 1,1,-1, 0,1,1, 0,0,1;
print(C);
```

6.1. HIGH LEVEL INTERFACE - USERS MANUAL

```
↦ 1,1,-1,
↦ 0,1,1,
↦ 0,0,1
def S = nc_algebra(C,0); setring S; S; // z is commutative
↦ //   characteristic : 0
↦ //   number of vars : 3
↦ //        block   1 : ordering dp
↦ //                : names    y(1) z y(2)
↦ //        block   2 : ordering C
↦ //   noncommutative relations:
↦ //     y(2)y(1)=-y(1)*y(2)
y(2)*y(1); // y(i) anti-commute
↦ -y(1)*y(2)
y(1)*y(1);
↦ y(1)^2
// Thus S is the correct anti-commutative algebra
ideal Q = y(1)^2, y(2)^2; // Squares
qring A = twostd(Q);
A;
↦ //   characteristic : 0
↦ //   number of vars : 3
↦ //        block   1 : ordering dp
↦ //                : names    y(1) z y(2)
↦ //        block   2 : ordering C
↦ //   noncommutative relations:
↦ //     y(2)y(1)=-y(1)*y(2)
↦ // quotient ring from ideal
↦ _[1]=y(2)^2
↦ _[2]=y(1)^2
y(2)*y(1);
↦ -y(1)*y(2)
y(1)*y(1); // not a zero, but:
↦ y(1)^2
NF(y(1)*y(1), std(0)); // zero in the quotient algebra
↦ 0
```

Given a commutative polynomial ring r, a super-commutative structure on it can be introduced using the SINGULAR' procedure SUPERCOMMUTATIVE from library nctools.lib (cf. [88]).

For example, the following quotient algebra of a graded commutative algebra:

$$(\mathbb{Q}[a,b] \otimes_{\mathbb{Q}} \wedge(x,y,z) \otimes_{\mathbb{Q}} \mathbb{Q}[Q,W]) / \langle a \otimes 1 \otimes W + b \otimes x \otimes Q + 1 \otimes z \otimes 1 \rangle$$

will be constructed in the following SINGULAR code:

```
LIB "nctools.lib";
ring r = 0,(a, b, x,y,z, Q, W),(lp(2), dp(3), Dp(2));
// Let us make variables x = var(3), ..., z = var(5) to be anti-commutative
// and add additionally a quotient ideal:
```

```
def A = superCommutative(3, 5, ideal(a*W + b*Q*x + z) ); setring A; A;
↦ //   characteristic : 0
↦ //   number of vars : 7
↦ //        block   1 : ordering lp
↦ //                : names    a b
↦ //        block   2 : ordering dp
↦ //                : names    x y z
↦ //        block   3 : ordering Dp
↦ //                : names    Q W
↦ //        block   4 : ordering C
↦ //   noncommutative relations:
↦ //     yx=-xy
↦ //     zx=-xz
↦ //     zy=-yz
↦ // quotient ring from ideal
↦ _[1]=z2
↦ _[2]=xz
↦ _[3]=y2
↦ _[4]=x2
↦ _[5]=bxyQ-yz
↦ _[6]=aW+bxQ+z
```

Note that unlike other non-commutative algebras, only non-commuting variables are required to be global (bigger than 1). In particular, commutative variables are allowed to be local, which means that one can deal with tensor products of any commutative rings with exterior algebras as described in Chapter 4. In the following SINGULAR example it would be the following tensor product:

$$\mathbb{Q}[a,b]_{\langle b \rangle} \otimes_\mathbb{Q} \wedge(x,y,z) \otimes_\mathbb{Q} \mathbb{Q}[Q,W]_{\langle Q,W \rangle}.$$

```
LIB "nctools.lib";
ring r = 0,(a, b, x,y,z, Q, W),(dp(1), ds(1), lp(3), ds(2));
def A = superCommutative(3, 5); setring A; A;
↦ //   characteristic : 0
↦ //   number of vars : 7
↦ //        block   1 : ordering dp
↦ //                : names    a
↦ //        block   2 : ordering ds
↦ //                : names    b
↦ //        block   3 : ordering lp
↦ //                : names    x y z
↦ //        block   4 : ordering ds
↦ //                : names    Q W
↦ //        block   5 : ordering C
↦ //   noncommutative relations:
↦ //     yx=-xy
↦ //     zx=-xz
↦ //     zy=-yz
```

6.1. HIGH LEVEL INTERFACE - USERS MANUAL

For user's convenience we also provide several graded commutative/super-commutative algebra-related procedures and functions in library `nctools.lib` (cf. [88]). Let us give detailed help and examples for them (cf. also SINGULAR Users Manual at [118]):

superCommutative

Usage:
   superCommutative([b,[e, [Q]]]);

Return:
   qring

Purpose:
   create a super-commutative algebra (as a GR-algebra) over a basering,

Note:
   activate this qring with the "setring" command.

Note:
   if b==e then the resulting ring is commutative.
   By default, `b=1, e=nvars(basering), Q=0`.

Theory:
   given a basering, this procedure introduces the anti-commutative relations
   $\text{var}(j)\text{var}(i)=-\text{var}(i)\text{var}(j)$ for all $e>=j>i>=b$ and creates the quotient
   of the anti-commutative algebra modulo the two-sided ideal, generated by
   $x(b)^2, ..., x(e)^2[+ Q]$

Display:
   If `printlevel > 1`, warning debug messages will be printed

Example:

```
LIB "nctools.lib";
ring R = 0,(x(1..4)),dp; // global!
def ER = superCommutative(); // the same as Exterior (b = 1, e = N)
setring ER; ER;
↦ //   characteristic : 0
↦ //   number of vars : 4
↦ //        block   1 : ordering dp
↦ //                : names    x(1) x(2) x(3) x(4)
↦ //        block   2 : ordering C
↦ //   noncommutative relations:
↦ //     x(2)x(1)=-x(1)*x(2)
↦ //     x(3)x(1)=-x(1)*x(3)
↦ //     x(4)x(1)=-x(1)*x(4)
```

```
↦ //     x(3)x(2)=-x(2)*x(3)
↦ //     x(4)x(2)=-x(2)*x(4)
↦ //     x(4)x(3)=-x(3)*x(4)
↦ // quotient ring from ideal
↦ _[1]=x(4)^2
↦ _[2]=x(3)^2
↦ _[3]=x(2)^2
↦ _[4]=x(1)^2
"Alternating variables: [", AltVarStart(), ",", AltVarEnd(), "].";
↦ Alternating variables: [ 1 , 4 ].
kill R; kill ER;
ring R = 0,(x(1..4)),(lp(1), dp(3)); // global!
def ER = superCommutative(2); // b = 2, e = N
setring ER; ER;
↦ //   characteristic : 0
↦ //   number of vars : 4
↦ //        block   1 : ordering lp
↦ //                : names    x(1)
↦ //        block   2 : ordering dp
↦ //                : names    x(2) x(3) x(4)
↦ //        block   3 : ordering C
↦ //   noncommutative relations:
↦ //     x(3)x(2)=-x(2)*x(3)
↦ //     x(4)x(2)=-x(2)*x(4)
↦ //     x(4)x(3)=-x(3)*x(4)
↦ // quotient ring from ideal
↦ _[1]=x(4)^2
↦ _[2]=x(3)^2
↦ _[3]=x(2)^2
"Alternating variables: [", AltVarStart(), ",", AltVarEnd(), "].";
↦ Alternating variables: [ 2 , 4 ].
kill R; kill ER;
ring R = 0,(x, y, z),(ds(1), dp(2)); // mixed!
def ER = superCommutative(2,3); // b = 2, e = 3
setring ER; ER;
↦ //   characteristic : 0
↦ //   number of vars : 3
↦ //        block   1 : ordering ds
↦ //                : names    x
↦ //        block   2 : ordering dp
↦ //                : names    y z
↦ //        block   3 : ordering C
↦ //   noncommutative relations:
↦ //     zy=-yz
↦ // quotient ring from ideal
↦ _[1]=y2
↦ _[2]=z2
"Alternating variables: [", AltVarStart(), ",", AltVarEnd(), "].";
↦ Alternating variables: [ 2 , 3 ].
x + 1 + z + y; // ordering on variables: y > z > 1 > x
↦ y+z+1+x
```

6.1. HIGH LEVEL INTERFACE - USERS MANUAL 91

```
std(x - x*x*x);
↦ _[1]=x
std(ideal(x - x*x*x, x*x*z + y, z + y*x*x));
↦ _[1]=y+x2z
↦ _[2]=z+x2y
↦ _[3]=x
kill R; kill ER;
ring R = 0,(x, y, z),(ds(1), dp(2)); // mixed!
def ER = superCommutative(2, 3, ideal(x - x*x, x*x*z + y, z + y*x*x )); // b = 2, e = 3
setring ER; ER;
↦ //   characteristic : 0
↦ //   number of vars : 3
↦ //        block   1 : ordering ds
↦ //                 : names    x
↦ //        block   2 : ordering dp
↦ //                 : names    y z
↦ //        block   3 : ordering C
↦ //   noncommutative relations:
↦ //     zy=-yz
↦ // quotient ring from ideal
↦ _[1]=y+x2z
↦ _[2]=z+x2y
↦ _[3]=x
↦ _[4]=y2
↦ _[5]=z2
"Alternating variables: [", AltVarStart(), ",", AltVarEnd(), "].";
↦ Alternating variables: [ 2 , 3 ].
```

IsSCA

Usage:
    IsSCA();

Return:
    int

Purpose:
    returns 1 if basering is a super-commutative algebra and 0 otherwise

Example:

```
LIB "nctools.lib";
////////////////////////////////////////////////////////////////
ring R = 0,(x(1..4)),dp; // commutative
if(IsSCA())
{ "Alternating variables: [", AltVarStart(), ",", AltVarEnd(), "]."; }
else
{ "Not a super-commutative algebra!!!"; }
```

92   CHAPTER 6. GRADED COMMUTATIVE ALGEBRAS IN SINGULAR

```
↦ Not a super-commutative algebra!!!
kill R;
////////////////////////////////////////////////////////////////
ring R = 0,(x(1..4)),dp;
def S = nc_algebra(1, 0); setring S; S; // still commutative!
↦ //   characteristic : 0
↦ //   number of vars : 4
↦ //        block   1 : ordering dp
↦ //                : names    x(1) x(2) x(3) x(4)
↦ //        block   2 : ordering C
↦ //   noncommutative relations:
if(IsSCA())
{ "Alternating variables: [", AltVarStart(), ",", AltVarEnd(), "]."; }
else
{ "Not a super-commutative algebra!!!"; }
↦ Not a super-commutative algebra!!!
kill R, S;
////////////////////////////////////////////////////////////////
ring R = 0,(x(1..4)),dp;
matrix E = UpOneMatrix(nvars(R));
int i, j; int b = 2; int e = 3;
for ( i = b; i < e; i++ )
{
  for ( j = i+1; j <= e; j++ )
  {
    E[i, j] = -1;
  }
}
def S = nc_algebra(E,0); setring S; S;
↦ //   characteristic : 0
↦ //   number of vars : 4
↦ //        block   1 : ordering dp
↦ //                : names    x(1) x(2) x(3) x(4)
↦ //        block   2 : ordering C
↦ //   noncommutative relations:
↦ //   x(3)x(2)=-x(2)*x(3)
if(IsSCA())
{ "Alternating variables: [", AltVarStart(), ",", AltVarEnd(), "]."; }
else
{ "Not a super-commutative algebra!!!"; }
↦ Not a super-commutative algebra!!!
kill R, S;
////////////////////////////////////////////////////////////////
ring R = 0,(x(1..4)),dp;
def ER = superCommutative(2); // (b = 2, e = N)
setring ER; ER;
↦ //   characteristic : 0
↦ //   number of vars : 4
↦ //        block   1 : ordering dp
↦ //                : names    x(1) x(2) x(3) x(4)
↦ //        block   2 : ordering C
```

6.1. HIGH LEVEL INTERFACE - USERS MANUAL   93

```
↦ //   noncommutative relations:
↦ //      x(3)x(2)=-x(2)*x(3)
↦ //      x(4)x(2)=-x(2)*x(4)
↦ //      x(4)x(3)=-x(3)*x(4)
↦ // quotient ring from ideal
↦ _[1]=x(4)^2
↦ _[2]=x(3)^2
↦ _[3]=x(2)^2
if(IsSCA())
{ "This is a SCA! Alternating variables: [", AltVarStart(), ",", AltVarEnd(), "]."; }
↦ This is a SCA! Alternating variables: [ 2 , 4 ].
```

AltVarStart

Usage:
 AltVarStart();

Return:
 int

Purpose:
 returns the number of the first alternating variable of basering

Note:
 basering should be a super-commutative algebra constructed by the procedure superCommutative, emits an error otherwise

Example:

```
LIB "nctools.lib";
ring R = 0,(x(1..4)),dp; // global!
"Alternating variables: [", AltVarStart(), ",", AltVarEnd(), "].";
↦     ? SCA rings are factors by (at least) squares!
↦     ? leaving nctools.lib::AltVarStart
def ER = superCommutative(2); // (b = 2, e = N)
setring ER; ER;
↦ //   characteristic : 0
↦ //   number of vars : 4
↦ //        block   1 : ordering dp
↦ //                  : names    x(1) x(2) x(3) x(4)
↦ //        block   2 : ordering C
↦ //   noncommutative relations:
↦ //      x(3)x(2)=-x(2)*x(3)
↦ //      x(4)x(2)=-x(2)*x(4)
↦ //      x(4)x(3)=-x(3)*x(4)
↦ // quotient ring from ideal
↦ _[1]=x(4)^2
```

```
⌐→ _[2]=x(3)^2
⌐→ _[3]=x(2)^2
"Alternating variables: [", AltVarStart(), ",", AltVarEnd(), "].";
⌐→ Alternating variables: [ 2 , 4 ].
setring R;
```

AltVarEnd

Usage:
   AltVarStart();

Return:
   int

Purpose:
   returns the number of the last alternating variable of basering

Note:
   basering should be a super-commutative algebra constructed by the procedure **superCommutative**, emits an error otherwise

Example:

```
LIB "nctools.lib";
ring R = 0,(x(1..4)),dp; // global!
"Alternating variables: [", AltVarStart(), ",", AltVarEnd(), "].";
⌐→    ? SCA rings are factors by (at least) squares!
⌐→    ? leaving nctools.lib::AltVarStart
def ER = superCommutative(2); // (b = 2, e = N)
setring ER; ER;
⌐→ //   characteristic : 0
⌐→ //   number of vars : 4
⌐→ //        block   1 : ordering dp
⌐→ //                 : names    x(1) x(2) x(3) x(4)
⌐→ //        block   2 : ordering C
⌐→ //   noncommutative relations:
⌐→ //     x(3)x(2)=-x(2)*x(3)
⌐→ //     x(4)x(2)=-x(2)*x(4)
⌐→ //     x(4)x(3)=-x(3)*x(4)
⌐→ // quotient ring from ideal
⌐→ _[1]=x(4)^2
⌐→ _[2]=x(3)^2
⌐→ _[3]=x(2)^2
"Alternating variables: [", AltVarStart(), ",", AltVarEnd(), "].";
⌐→ Alternating variables: [ 2 , 4 ].
```

6.2 Product of monomials in graded commutative algebras

The formula for computing the sign change in a antic-commutative algebra is very simple:

$$(\xi_1^{\alpha_1} \cdots \xi_m^{\alpha_m}) * (\xi_1^{\beta_1} \cdots \xi_m^{\beta_m}) = (-1)^\sigma \xi_1^{\alpha_1+\beta_1} \cdots \xi_m^{\alpha_m+\beta_m},$$

where $\sigma = \sum_{i=1}^{m-1} \sum_{j=i+1}^{m} \beta_i \alpha_j$ (mod 2). Observe that since squares are zeroes in a graded commutative algebra the product is nonzero only if all sums $\alpha_i + \beta_i$ are smaller than 2.

The following SINGULAR-kernel C/C++ procedure shows how to compute σ in one go:

```
/// returns the sign (+1/-1) of lm(pMonomA) * lm(pMonomB)
/// returns 0 if the product is zero
/// preserves input polynomials
int sca_Sign_mm_Mult_mm( poly pMonomA, poly pMonomB, ring rRing )
{
  unsigned int sigma = 0; // resulting power of -1
  unsigned int alphasum = 0; // intermediate power
  // process both exponents in one go:
  int j  = scaLastAltVar(rRing);
  while( j != scaFirstAltVar(rRing) )
  {
    // get j-th exponents of both leading monomials:
    unsigned int iA = p_GetExp(pMonomA, j, rRing); // A's,
    unsigned int iB = p_GetExp(pMonomB, j, rRing); // B's

    if( iB != 0 )
    {
      if( iA != 0 ) // same anti-comm. var in both exponents?
        return (0);

      sigma ^= alphasum; // sigma += iB * alphasum (mod 2)
    }

    alphasum ^= iA; // alphasum += iA (mod 2)
    j = j - 1;
  }
  return(1 - 2*sigma); // 1 is odd => -1, 0 is even => 1
}
```

Due to the simplicity of computing a product of monomials (see above) there is no need in any caching, which is always done by the general non-commutative implementation from SINGULAR:PLURAL.

6.3 Detection of a graded commutative structure

Conceptually, graded commutative algebras are introduced and defined in SINGULAR as factors of anti-commutative algebras by two-sided ideals containing squares of all anti-commuting variables, i.e. as GR-algebras.

Remark 6.3.1. Due to performance penalty we will not consider commutative algebras to be graded commutative. Therefore our implementation awaits at least two anti-commuting variables.

Therefore we first need to find all variables which anti-commute and verify that all other variables are commutative. Next in order to check the factor ideal we simply verify that squares of all anti-commuting variables reduce to zero module factor ideal, which is given by a two-sided GB.

The first check means that the upper triangular matrix C from the definition of GR-algebras has only ± 1, and $C_{i,j} = -1$ iff both i-th and j-th variables are anti-commuting. That is we can reorder variables so that C have the following block form:

where A and B are square upper uni-triangular matrices (entries on the main diagonal are 1) with the entries above the main diagonal being all -1 for A and respectively all 1 for B.

Therefore our detection procedure goes as follows:

1. find topmost row in C containing at least one entry of -1,
2. find all such entries in that row, let us say they have indices $i_1, \ldots, i_k$,
3. verify that all corresponding variables anti-commute, that is, $C_{i,j} = -1$ for all $j, i \in \{i_1, \ldots, i_k\} : j > i$,
4. all other entries in C must be 1,
5. check whether squares of found anti-commuting variables are contained in the quotient ideal.

6.4 A bit about SINGULAR internals

Let us give a brief glance over the SINGULAR internal structure[1]. The following diagram shows some of its main "components" as well as our non-commutative subsystems: **Singular** – interpreter shell, **kernel** – mathematical kernel, **Plural** – non-commutative kernel

---
[1]Some more documentation is available online at http://www.mathematik.uni-kl.de/ftp/pub/Math/Singular/doc/, in Section 6.5

extension, **SCA** – subsystem of SINGULAR:PLURAL which deals with graded commutative algebras.

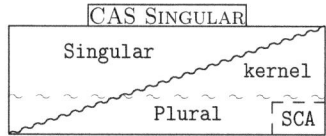

A newbie wanting to program (or even read and understand) SINGULAR kernel faces the problem that there is no really strict decomposition into separated and independent components. Moreover, some functions rely on the knowledge about implementations, data structures and even implicitly assumed assumptions. That is from computer science point of view the CAS SINGULAR has high coupling and weak cohesion with all the consequences. For instance the above diagram shows that Singular and kernel are not independent, e.g. portions of kernel functionality are inserted directly into the interpreter but the non-commutative kernel extension Plural is completely separated from the interpreter and has clear a interface with the kernel. In turn Plural has a transparent interface for special non-commutative extensions, e.g. SCA.

The main object in SINGULAR is ring. It defines the structure of data objects called poly. A poly corresponds to a vector (resp. polynomial) in a module (resp. ideal) over the commutative or non-commutative polynomial or quotient algebras over a field or number ring, described by corresponding ring. Internally (cf. Section 6.5 and references thereof) a poly is a sorted one-directional list of terms, whereas every terms contains a coefficient c, pointer to the next terms and (embedded) exponent array, which encodes an exponent α and module component i of some module term: $cx^\alpha \epsilon_i$.

While non-commutative algebras require more information than commutative ones, their elements such as (commutative and non-commutative) polynomials and vectors are internally the same.

Each object of type ring has a table of function pointers (e.g. virtual methods) for overloading basic arithmetic operations with vectors (i.e. poly) and coefficients.

For the non-commutative case we have extended the set of basic operation as well as some higher level operation (S-polynomial, GB-computation).

6.5 Implementing an induced ordering after Schreyer

The existing SINGULAR algorithms for the computation of Schreyer resolution (sres requires a GB) and La Scala resolution (lres can start with homogeneous ideals only) work over commutative polynomial algebras only. Moreover they simulate the Schreyer ordering by working directly with induced products of monomials needed to compare module terms with respect to Schreyer ordering. This results in a lot of transformations whenever the actual leading monomials are needed, which makes reductions especially expensive.

Recently there appeared some new promising improvements (cf. [47, 20] etc.) for the computation of GBs, which deal with some kind of syzygies, typically making use of the Schreyer ordering. In order to be able to test and compare them in SINGULAR and implement our syzygy-driven algorithms proposed in Chapter 5, we need the Schreyer ordering (due to Section 5.3), which would be accessible from the interpreter script language and be transparent for the users.

In order to achieve high performance in GB-related computations SINGULAR has a very sophisticated mechanism for constructing monomials. The key design idea is the following: monomial ordering defines the low-level structure of monomial exponents (cf. [8], online documents available at http://www.mathematik.uni-kl.de/ftp/pub/Math/Singular/doc/ and http://convex-singular.googlecode.com/files/singular.pdf).

That is why any new monomial ordering has to be implemented deep in the SINGULAR kernel.

Note that SINGULAR stores exponent as vectors of integers, which are interpreted via a set of rules, as described by the corresponding ring (cf [8]).

The main design requirement is that exponent arrays must be additive, i.e. multiplication with a monomial amounts to summing up exponents, without any additional computations. Let's inspect the structure of a monomial in $P = \mathbb{Q}[x, y, z]$ with respect to <degrevlex,C> on a 64-bit Linux PC:

```
ring P = (0), (x, y, z), (dp, C); // define the ring P
system("DetailedPrint", P); // print details about P
↦ //   characteristic : 0
↦ //   number of vars : 3
↦ //        block   1 : ordering dp
↦ //                  : names    x y z
↦ //        block   2 : ordering C
↦ ExpL_Size:3 CmpL_Size:3 VarL_Size:1
↦ bitmask=0xffffff (expbound=1048575)
↦ BitsPerExp=20 ExpPerLong=3 MinExpPerLong=3 at L[1]
↦ varoffset:
↦   v0 at e-pos 2, bit 0
↦   v1 at e-pos 1, bit 0
↦   v2 at e-pos 1, bit 20
↦   v3 at e-pos 1, bit 40
↦ divmask=1000010000100001
↦ ordsgn:
↦   ordsgn 1 at pos 0
↦   ordsgn -1 at pos 1
↦   ordsgn 1 at pos 2
↦ OrdSgn:1
```

6.5. IMPLEMENTING AN INDUCED ORDERING AFTER SCHREYER

```
↦ ordrec:
↦    typ ro_dp  place 0  start 1  end 3
↦ pOrdIndex:0 pCompIndex:2
↦ OrdSize:1
↦ --------------------
↦ L[0]: ordsgn 1 ordrec:ro_dp (start:1, end:3) pOrdIndex
↦ L[1]: ordsgn -1 v1 at e[1], bit 0; v2 at e[1], bit 20; v3 at e[1], bit 40;
↦ L[2]: ordsgn 1 v0;
↦    [...omitted...]
system("DetailedPrint", x * gen(2));
↦ x*gen(2)
↦
↦ exp[0..2]
↦ 000000001 000000001 000000002
↦ v0:        2  v1:     1 v2:     0 v3:     0
```

It shows that monomial exponents have the following structure: first goes the degree next comes a single machine integer comprising all three variable powers and at last – the module component:

$$\begin{array}{|c|c|c|} \hline deg & \{-z, -y, -x\} & c \\ \hline \end{array}$$

This packed (vectorized) structure allows SINGULAR to compare monomials via a simple lexicographical ordering on vectors of integers.

In our implementation of Schreyer induced orderings we model an elementary arrow $(g \rightsquigarrow f)$ by the sum $f + g$ living in somewhat bigger free module. Moreover we have to ensure that all terms from g are stored at the end of the sum (e.g. a monomial from g has to be smaller than monomials from f.

Moreover, in order to achieve this and also due to Section 5.4 we add a level marker at the beginning of the comparable part of monomial exponent and ensure that monomials from with smaller level are bigger.

By using such an implementation for elementary arrows we also achieve that $LT(g \rightsquigarrow f) = $ Lt(f) as needed by our definition of an elementary arrow

Now we sketch our implementation of classical Schreyer induced orderings in SINGULAR. Our interface consists of two functions: MakeInducedSchreyerOrdering (which creates a ring capable of treating Schreyer induced ordering) and SetInducedReferrence (which activates it by setting the reference information).

```
def S = system("MakeInducedSchreyerOrdering"); // initialization
setring S; system("DetailedPrint", S); // print details about S
↦ //   characteristic : 0
↦ //   number of vars : 3
↦ //        block   1 : ordering IS::prefix
↦ //        block   2 : ordering dp
↦ //                  : names    x y z
```

```
↦ //        block    3 : ordering C
↦ //        block    4 : ordering IS::suffix (sign: 1)
↦ ExpL_Size:6 CmpL_Size:5 VarL_Size:1
↦ bitmask=0xfffff (expbound=1048575)
↦ BitsPerExp=20 ExpPerLong=3 MinExpPerLong=3 at L[5]
↦ varoffset:
↦   v0 at e-pos 4, bit 0
↦   v1 at e-pos 5, bit 40
↦   v2 at e-pos 5, bit 20
↦   v3 at e-pos 5, bit 0
↦ divmask=1000010000100001
↦ ordsgn:
↦   ordsgn -1 at pos 0
↦   ordsgn 1 at pos 1
↦   ordsgn -1 at pos 2
↦   ordsgn 1 at pos 3
↦   ordsgn 1 at pos 4
↦ OrdSgn:1
↦ ordrec:
↦   typ ro_isTemp  start (level) 0, suffixpos: 2, V0:
↦   typ ro_dp  place 1  start 1  end 3
↦   typ ro_is  start 0, end: 3:   limit 0
↦   F: (NULL)weights: NULL == [0,...0]
↦
↦ pOrdIndex:1 pCompIndex:4
↦ OrdSize:3
↦ --------------------
↦ L[0]: ordsgn -1 ordrec:ro_isTemp (start:0, end:2) ordrec:ro_is (start:3, end:0)
↦ L[1]: ordsgn 1 ordrec:ro_dp (start:1, end:3) pOrdIndex
↦ L[2]: ordsgn -1
↦ L[3]: ordsgn 1
↦ L[4]: ordsgn 1 v0;
↦ L[5]: no comp v1 at e[5], bit 40; v2 at e[5], bit 20; v3 at e[5], bit 0;
↦   [...omitted...]
```

In order to setup an induced ordering we first need a set of module terms $m_1, \ldots, m_k$ of rank at most r, with respect to which we will setup the Schreyer induced ordering as follows:

```
module M = m_1, ..., m_k;
system("SetInducedReferrence", M, 2); // Setup Schreyer ordering on S wrt M
```

By using `SetInducedReferrence` iteratively we can easily construct the Schreyer frame and perform needed setup for our free resolution algorithm (cf. Algorithm 5.4.4).

In this case induced exponents will look as follows:

$-level$	deg'	$\{-z', -y', -x'\}$	c'	c	$\overline{\{x, y, z\}}$

6.5. IMPLEMENTING AN INDUCED ORDERING AFTER SCHREYER

First comes the level marker needed to compare module terms coming from different levels (free modules). Next comes the "induced" part (3 integers, primed on the above figure) which is hidden from the user and simply allows SINGULAR to compare induced products (saved here verbatim as in the original ring), including the original module component on the very first level. The last comparable entry is the actual module component (or alternatively its negative, depending on the needed Schreyer ordering). Only these first 5 integers are to be compared. The last additional "exponent" comprises the actual variable powers for fast look-up .

Unfortunately, this nested (memory hungry) structure is the only way to avoid repeated expensive intermediate transformations (as implemented in sres and lres), needed for fast computation of leading information and reductions, and allow fast comparement w.r.t. Schreyer ordering.

This way the module component ϵ_{r+i} to have the same induced part as m_i (and bigger level). For instance: if $M = (x \cdot \epsilon_2)$ then $x \cdot \epsilon_2$ and ϵ_2 have the following internal representations :

```
module M = x * gen(2);
system("SetInducedReferrence", M, 2); // Setup Schreyer ordering on S wrt M
system("DetailedPrint", S); // S has been changed
↦ //   characteristic : 0
↦ //   number of vars : 3
↦ //        block   1 : ordering IS::prefix
↦ //        block   2 : ordering dp
↦ //                  : names    x y z
↦ //        block   3 : ordering C
↦ //        block   4 : ordering IS::suffix (sign: 1)
↦ ExpL_Size:6 CmpL_Size:5 VarL_Size:1
↦ bitmask=0xffffff (expbound=1048575)
↦ BitsPerExp=20 ExpPerLong=3 MinExpPerLong=3 at L[5]
↦ varoffset:
↦    v0 at e-pos 4, bit 0
↦    v1 at e-pos 5, bit 40
↦    v2 at e-pos 5, bit 20
↦    v3 at e-pos 5, bit 0
↦ divmask=1000010000100001
↦ ordsgn:
↦    ordsgn -1 at pos 0
↦    ordsgn  1 at pos 1
↦    ordsgn -1 at pos 2
↦    ordsgn  1 at pos 3
↦    ordsgn  1 at pos 4
↦ OrdSgn:1
↦ ordrec:
↦    typ ro_isTemp  start (level) 0, suffixpos: 2, V0:
↦    typ ro_dp   place 1  start 1  end 3
↦    typ ro_is   start 0, end: 3:   limit 2
↦    F: Module of rank 2,real rank 2 and 1 generators.
↦ generator 0: x*gen(2)
```

```
↦
↦ exp[0..5]
↦ 000000001 000000001 000000010 000000002 000000002 10000000000
↦ v0:       2  v1:   1 v2:   0 v3:   0
↦ weights: NULL == [0,...0]
↦
↦ pOrdIndex:1 pCompIndex:4
↦ OrdSize:3
↦ --------------------
↦ L[0]: ordsgn -1 ordrec:ro_isTemp (start:0, end:2) ordrec:ro_is (start:3, end:0)
↦ L[1]: ordsgn 1 ordrec:ro_dp (start:1, end:3) pOrdIndex
↦ L[2]: ordsgn -1
↦ L[3]: ordsgn 1
↦ L[4]: ordsgn 1 v0;
↦ L[5]: no comp v1 at e[5], bit 40; v2 at e[5], bit 20; v3 at e[5], bit 0;
↦ [...omitted...]
system("DetailedPrint", gen(3) + x*gen(2)); // print details about it
↦ x*gen(2)+gen(3)
↦
↦ exp[0..5]
↦ 000000001 000000001 000000010 000000002 000000002 10000000000
↦ v0:       2  v1:   1 v2:   0 v3:   0
↦
↦ exp[0..5]
↦ 000000002 000000001 000000010 000000002 000000003 000000000
↦ v0:       3  v1:   0 v2:   0 v3:   0
```

Chapter 7

Applications

In this chapter we illustrate some applications of the developed framework in Projective Geometry, Affine Geometry and Physics.

7.1 Projective Geometry

This section is devoted to the computation of sheaf cohomology of coherent sheaves over a projective variety. Please refer to the classical texts [73, 115] and [57] for basic definitions. An excellent introduction to computational methods was given by M. Stillman at the Arizona Winter School in Tucson, March 2006, (notes [122] and videos are available online at http://math.arizona.edu/~swc/aws/06/06Notes.html). Another short overview can be found in [30].

Traditional methods for computing with sheaves and sheaf cohomology (shortly described in Section 7.1.1) can be found in [115], [126] and [119].

The exterior algebra method for computing sheaf cohomology (described in Section 7.1.2) relies on a constructive version of BGG correspondence given in [36] and [29]. Note that the BGG correspondence (cf. [15, 11]) is a particular case of Koszul duality (cf. [53, 52, 12]).

The implementation of this method by W. Decker, D. Eisenbud and F.-O. Schreyer as package `BGG` (cf. [29]) in M2 (cf. [61, 126]) was the fastest known up until now.

Generalization of this method allows one to compute higher direct images of sheaves (cf. [39]). Apart from constructing Beilinson Monad and Horrocks-Mumford bundle, this approach has been used for plenty interesting theoretical applications, e.g. compution of resultants and Chow forms (cf. [35, 42]), cohomologies of hyperplane arrangement (cf. [38]), and others (e.g. [43]).

This method and most of its applications, for example the investigation of the *Minimal Resolution Conjecture* (cf. [37]), require complicated computations. Therefore, it is important to have a robust (practically efficient) Computer Algebra framework, which would support this method and further experimental research (e.g. due to [40] and [41]).

7.1.1 Introduction to sheaf cohomology

?

Let $S := \Bbbk[x_0, \ldots, x_n]$ denote a homogeneous coordinate ring of $\mathbb{P}^n$ over a field $\Bbbk$ (not necessarily algebraically closed) and ($\mathbb{Z}$-)graded by putting $\deg x_i = 1$. Let I be a homogeneous ideal in S, $X = \mathrm{V}(I) \subset \mathbb{P}^n$ a *projective variety* with (homogeneous) coordinate ring $R = S/I$.

Sheaves on X organize local data on X. For our "chosen" definition of a coherent sheaf on $\mathbb{P}^n$ see Remark 7.1.8.

We denote $\mathcal{O}(i) := \mathcal{O}_X(i)$ – the twisted invertible sheaves of Serre on X (in particular, $\mathcal{O} = \mathcal{O}_X(0)$ is the structure sheaf) and $\mathcal{F}(i) := \mathcal{F} \otimes_\mathcal{O} \mathcal{O}(i)$ – the i-th twist of a coherent sheaf $\mathcal{F}$ on X.

Cohomology functors $\mathrm{H}^j(X, -)$ may be defined as right derived functors of the global sections functor: $\mathrm{H}^0(X, -) = \Gamma(X, -)$ similary to the functors $\mathrm{Ext}^i(M, -)$ being the right derived of $\mathrm{Hom}(M, -)$ or the *higher direct image* functors $R^i \pi_*$ being the right derived functor of π_* for a morphism of schemes $X \xrightarrow{\pi} Y$.

Let us denote and the j-th cohomology group of $\mathcal{F}(i)$ by $\mathrm{H}^j \mathcal{F}(i) := \mathrm{H}^j(X, \mathcal{F}(i))$. Given a coherent sheaf $\mathcal{F}$ on X, computing its cohomology means computing one of the following:

- one of the dimensions $\mathrm{h}^j \mathcal{F}(i) := \dim_\Bbbk \mathrm{H}^j \mathcal{F}(i)$.
- these dimensions in a certain range of twists.
- the graded R-modules (called *cohomology modules*)

$$\mathrm{H}^j_* \mathcal{F} := \mathrm{H}^j_*(X, \mathcal{F}) := \bigoplus_{i \in \mathbb{Z}} \mathrm{H}^j \mathcal{F}(i)$$

  measure how complicated the sheaf $\mathcal{F}$ is.
- the truncation of $\mathrm{H}^j_* \mathcal{F}$ at degree $d \in \mathbb{Z}$:

$$\mathrm{H}^j_{\geq d}(X, \mathcal{F}) := \bigoplus_{i \geq d} \mathrm{H}^j \mathcal{F}(i)$$

Remark 7.1.1. We shall usually display cohomology dimensions in the form of the following **cohomology table**:

$$\begin{matrix} \mathrm{h}^n \mathcal{F}(l) & \mathrm{h}^n \mathcal{F}(l+1) & \ldots & \mathrm{h}^n \mathcal{F}(h) \\ \vdots & \vdots & \ddots & \vdots \\ \mathrm{h}^0 \mathcal{F}(l) & \mathrm{h}^0 \mathcal{F}(l+1) & \ldots & \mathrm{h}^0 \mathcal{F}(h) \end{matrix},$$

for two given bounds $l, h \in \mathbb{Z}$ such that $l \leq h$.

Note that the we will only consider the projective case since affine case is not interesting due to the following remark:

7.1. PROJECTIVE GEOMETRY

Remark 7.1.2 (Chapter III, Theorem 3.5, Remark 3.5.1 in [73]). Let $X = \operatorname{Spec} A$ be an *affine variety*, where A is a commutative ring, then for any quasi-coherent sheaves $\mathcal{F}$ on X, and all $j > 0$: $\operatorname{H}^j(X, \mathcal{F}) = 0$.

An axiomatic approach (e.g. due to Čech) defines sheaf cohomology using long exact sequences and its values on specific sheaves: let in our setting $\{U_i \mid 0 \leqslant i \leqslant k\}$ be an open affine cover of X, the standard one is defined by the affine open sets $U_i = X \setminus \operatorname{V}(x_i) \subset \mathbb{P}^n$ in X (in particular $k = n$). Now let for $\lambda = \{\lambda_0, \ldots, \lambda_p\} \subset \{0, \ldots, k\} : |\lambda| := p, U_\lambda := \bigcap_{i=0}^{p} U_{\lambda_i}$.

Definition 7.1.3 (Čech complex). For any $p \in \{0, \ldots, k\}$ let $\mathcal{C}^p(\mathcal{F}) := \bigoplus_{|\lambda|=p} \mathcal{F}(U_\lambda)$ and the natural map $\sigma_p : \mathcal{C}^p(\mathcal{F}) \to \mathcal{C}^{p+1}(\mathcal{F}) : f_{i_0, \ldots, i_p} \mapsto g_{j_0, \ldots, j_{p+1}}$, where $g_{j_0, \ldots, j_{p+1}} := \sum_{i=0}^{p+1} (-1)^i f_{j_0, \ldots, \widehat{j_i}, \ldots, j_{p+1}}$, and $\widehat{j_i}$ means that this index is omitted.

The **Čech complex** of $\mathcal{F}$ is the complex (of infinite dimensional vector spaces over $\Bbbk$):

$$0 \to \mathcal{C}^0(\mathcal{F}) \xrightarrow{\sigma_0} \mathcal{C}^1(\mathcal{F}) \xrightarrow{\sigma_1} \cdots \to \mathcal{C}^k(\mathcal{F}) \to 0 \tag{7.1}$$

Let now $\operatorname{H}^j(\mathcal{F}) = \operatorname{H}^j(X, \mathcal{F})$ be the j-th cohomology of the Čech complex (7.1). Choosing another affine open cover will yield isomorphic cohomology groups.

Any sheaf $\mathcal{F}$ defined on a projective subvariety or subscheme $X \subset \mathbb{P}^n$ with the use of the inclusion map $i : X \to \mathbb{P}^n$ can be thought of as a sheaf $i_* \mathcal{F}$ on $\mathbb{P}^n$ which coincides with $\mathcal{F}$ on X and is the 0-sheaf outside of X, and has the same cohomologies: $\operatorname{H}^j(X, \mathcal{F}) \cong \operatorname{H}^j(\mathbb{P}^n, i_* \mathcal{F})$.

Proposition 7.1.4 (Cohomology properties due to [115]). *Let $\mathcal{F}$ be a coherent sheaf on $\mathbb{P}^n$. Then*

- *There exists d_0 such that $\mathcal{F}(d)$ is generated by finitely many global sections for all $d \geqslant d_0$.*
- *For all $0 \leqslant j \leqslant n$: $\operatorname{H}^j(\mathcal{F})$ is a finite dimensional vector space over $\Bbbk$.*
- *$\operatorname{H}^j(\mathcal{F}) = 0$ for all $j > d$, where $d := \dim \operatorname{supp} \mathcal{F}$.*
- *$\operatorname{H}^j(\mathcal{F}(d)) = 0$ for all $j > 0$ and $d \gg 0$.*
- *Let*

$$0 \to \mathcal{F}' \to \mathcal{F} \to \mathcal{F}'' \to 0$$

be a short exact sequence of coherent sheaves on $\mathbb{P}^n$. Then there exist connecting homomorphisms $\delta_j : \operatorname{H}^j(\mathcal{F}'') \to \operatorname{H}^{j+1}(\mathcal{F}')$ such that the following long sequence is exact:

$$0 \to \operatorname{H}^0(\mathcal{F}') \to \operatorname{H}^0(\mathcal{F}) \to \operatorname{H}^0(\mathcal{F}'') \xrightarrow{\delta_0} \operatorname{H}^1(\mathcal{F}') \to \operatorname{H}^1(\mathcal{F}) \to \cdots$$

Example 7.1.5 (cf. [115]). Let $S = \Bbbk[x_0, \ldots, x_n], d \in \mathbb{Z}$ then it follows from the above:

$$\operatorname{H}^0(\mathbb{P}^n, \mathcal{O}_{\mathbb{P}^n}(d)) = S_d, \operatorname{H}^i(\mathbb{P}^n, \mathcal{O}_{\mathbb{P}^n}(d)) = 0, \operatorname{H}^n(\mathbb{P}^n, \mathcal{O}_{\mathbb{P}^n}(d)) = S^*_{-n-1-d},$$

where $0 < i < n$ and the star (*) denotes the k-vector space dual.

For example computations see the SINGULAR procedures sheafCoh and sheafCohBGG, described in this and the next sections.

We shall use Serre's sheafification functor $M \mapsto \widetilde{M}$ (cf. [115]) in order to represent a coherent sheaf $\mathcal{F}$ on $\mathbb{P}^n$ by a ($\mathbb{Z}$-)graded S-module M, which is either finitely generated itself, or "eventually finitely generated", i.e. some truncation $M_{\geqslant d}$ is finitely generated. Moreover we identify two modules N and N' iff there exists r: $N_{\geqslant r} \cong N'_{\geqslant r}$ since they correspond to isomorphic coherent sheaves on $\mathbb{P}^n$.

Remark 7.1.6 (Construction of $\widetilde{M}$). Let M be a graded S-module, which is eventually finitely generated, we associate to it the coherent sheaf of $\mathcal{O}_{\mathbb{P}^n}$-modules on $\mathbb{P}^n$ denoted by $\widetilde{M}$, defined (due to Ex. II.1.23 in [73]) by gluing the sheaves:

$$\widetilde{M}(U_i) := \left(M \otimes_S S[x_i^{-1}]\right)_0, \left(\widetilde{M}|_{U_i}\right)_{U_i \cap U_j} = \left(M \otimes_S S[x_i^{-1}] \otimes_S S[x_j^{-1}]\right)_0$$

via the maps

$$\left(M \otimes_S S[x_i^{-1}] \otimes_S S[x_j^{-1}]\right)_0 \to \left(M \otimes_S S[x_j^{-1}] \otimes_S S[x_i^{-1}]\right)_0 :$$
$$m \otimes f \otimes g \mapsto m \otimes g \otimes f,$$

where the 0 subindex denotes the subset of all elements of degree 0 and $\{U_i \mid 0 \leqslant i \leqslant n\}$ is the standard open affine cover of $\mathbb{P}^n$, with affine intersections.

Note that $\widetilde{M}(U_i)$ is a finitely generated $\mathcal{O}(U_i)$-module due to our assumptions on M.

Remark 7.1.7. Similary we could have defined a sheaf $\widetilde{M}$ on X, for $X = V(I) \subset \mathbb{P}^n$ and a finitely generated graded S/I-module M.

Moreover there is also a dual functor which maps a coherent $\mathcal{O}_{\mathbb{P}^n}$ module $\mathcal{F}$ to an eventually finitely generated graded S-module $H^0_*(\mathbb{P}^n, \mathcal{F}) = \bigoplus_{d \in \mathbb{Z}} H^0(\mathbb{P}^n, \mathcal{F}(d))$, which is also a graded S/I-module if $\mathcal{F}$ is the extension by zero of a sheaf on $X = VI \subset \mathbb{P}^n$.

The above constructions have the following properties:

1. $\widetilde{S(d)} = \mathcal{O}_{\mathbb{P}^n}(d)$, in particular, $\widetilde{S} = \mathcal{O}_{\mathbb{P}^n}$.
2. $\widetilde{M}(d) = \widetilde{M(d)}$
3. If I is an ideal in S defining subvariety $X \subset \mathbb{P}^n$ then $\widetilde{I} = \mathcal{J}_X$ (the ideal sheaf on $\mathbb{P}^n$, i.e. local equations of X). In particular, if $R = S/I$ then $\mathcal{O}_X(d) = \widetilde{R(d)}$.
   If moreover $D \subset X$ is a subvariety of comdim. 1 with ideal $J \subset S/I$ then $\mathcal{O}_X(-D) = \widetilde{J}$.
4. Let $X \subset \mathbb{P}^n$ defined by ideal $I \subset S$ then the *normal bundle (sheaf)* $N_{X/\mathbb{P}^n} = \widetilde{\text{Hom}_S(I, S/I)}$.
5. If M is an eventually finitely generated graded S-module such that $M_d = 0$ for all $d \gg 0$, then $\widetilde{M} = 0$.

7.1. PROJECTIVE GEOMETRY

6. Every coherent $\mathcal{O}_{\mathbb{P}^n}$-module is isomorphic to $\widetilde{M}$, for some finitely generated graded S-module M.
7. For every coherent $\mathcal{O}$-module $\mathcal{F}$ holds $\widetilde{\mathrm{H}^0_*(\mathbb{P}^n, \mathcal{F})} = \mathcal{F}$.
8. If M is a graded S-module, then the natural map $M \to \mathrm{H}^0_*\widetilde{M}$
9. The operation $M \mapsto \widetilde{M}$ is an exact functor from the category of eventually finitely generated graded S-modules to the category of coherent $\mathcal{O}_{\mathbb{P}^n}$-modules (denoted by $\mathrm{Coh}\mathbb{P}^n$)
10. The functors $\widetilde{(-)}$ and $\mathrm{H}^0_*(-)$ provide an equivalence of categories between $\mathrm{Coh}\mathbb{P}^n$ and the category of eventually finitely generated S-modules, but the functor $\mathrm{H}^0_*(-)$ is not exact.

Remark 7.1.8. For our needs Part 6 of Remark 7.1.6 can be used as a definition of a coherent sheaf on $\mathbb{P}^n$.

From now on, we denote $\mathcal{O}(i) := \mathcal{O}_{\mathbb{P}^n}(i)$ the line bundles on $\mathbb{P}^n$ (in particular, $\mathcal{O} = \mathcal{O}(0)$ is the structure sheaf), a coherent sheaf $\mathcal{F} = \widetilde{M}$ on $\mathbb{P}^n$, for some finitely generated graded S-module M. The i-th twist of $\mathcal{F}$ is denoted by $\mathcal{F}(i) := \mathcal{F} \otimes_{\mathcal{O}} \mathcal{O}(i)$, and the j-th cohomology group of $\mathcal{F}$ by $\mathrm{H}^j \mathcal{F} = \mathrm{H}^j(\mathbb{P}^n, \mathcal{F})$. Then

$$\mathrm{H}^j_* \mathcal{F} := \bigoplus_{i \in \mathbb{Z}} \mathrm{H}^j \mathcal{F}(i)$$

is a graded S-module.

Remark 7.1.9. Following [126, Chapter 8] one can compute sheaf cohomology due to the *local duality*:

- For all $j \geqslant 1$:
$$\mathrm{H}^j_* \widetilde{M} \cong \mathrm{Ext}^{n-j}_S(M, S(-n-1))^\vee,$$
- And the following exact sequence in the case $j = 0$:
$$0 \to \mathrm{Ext}^{n+1}_S(M, S(-n-1))^\vee \to M \to \mathrm{H}^0_* \widetilde{M} \to \mathrm{Ext}^n_S(M, S(-n-1))^\vee \to 0,$$
where the graded vector space dual $\left(\bigoplus_{i \in \mathbb{Z}} M_i\right)^\vee = \bigoplus_{i \in \mathbb{Z}} \mathrm{Hom}_{\Bbbk}(M_{-i}, \Bbbk)$. is endowed with its natural structure as a graded S-module.

It follows that
$$\mathrm{h}^j \widetilde{M}(d) = \dim_{\Bbbk} \mathrm{Ext}^{n-j}_S(M, S)_{-n-1-d},$$
and the dimension of the space of global sections of $\widetilde{M}(d)$:
$$\mathrm{h}^0 \widetilde{M}(d) = \dim_{\Bbbk} M_d + \dim_{\Bbbk} \mathrm{Ext}^n_S(M, S)_{-n-1-d} - \dim_{\Bbbk} \mathrm{Ext}^{n+1}_S(M, S)_{-n-1-d}.$$

The method for computing sheaf cohomology relies, because of the above local duality isomorphisms, on computing free resolutions over S for computing Ext. This method has been implemented in SINGULAR (cf. command `sheafCoh` from `sheafcoh.lib`).

sheafCoh

Usage:
sheafCoh(M,l,h); M module, l,h int

Assume:
M is graded, and it comes assigned with an admissible degree vector as an attribute, h>=1. The basering S has n+1 variables.

Return:
intmat, cohomology of twists of the coherent sheaf F on P^n associated to coker(M). The range of twists is determined by l, h.

Display:
The intmat is displayed in a diagram of the following form:

	l	l+1		h
n:	h^n(F(l))	h^n(F(l+1))	......	h^n(F(h))
	....	....	......	....
1:	h^1(F(l))	h^1(F(l+1))	......	h^1(F(h))
0:	h^0(F(l))	h^0(F(l+1))	......	h^0(F(h))
chi:	chi(F(l))	chi(F(l+1))	......	chi(F(h))

A '-' in the diagram refers to a zero entry.

Note:
The procedure is based on local duality as described in [Eisenbud: Computing cohomology. In Vasconcelos: Computational methods in commutative algebra and algebraic geometry. Springer (1998)].

By default, the procedure uses mres to compute the Ext modules. If called with the additional parameter "sres", the sres command is used instead.

Example:

```
LIB "sheafcoh.lib";
//
// cohomology of structure sheaf on P^4:
//
ring r=0,x(1..5),dp;
module M=0; // Corresponds to the structure sheaf on Proj(r)=P^4
def A=sheafCoh(M,-7,2);
```

7.1. PROJECTIVE GEOMETRY

```
↦          -7  -6  -5  -4  -3  -2  -1   0   1   2
↦        ----------------------------------------
↦     4:  15   5   1   -   -   -   -   -   -   -
↦     3:   -   -   -   -   -   -   -   -   -   -
↦     2:   -   -   -   -   -   -   -   -   -   -
↦     1:   -   -   -   -   -   -   -   -   -   -
↦     0:   -   -   -   -   -   -   -   1   5  15
↦        ----------------------------------------
↦  chi:  15   5   1   0   0   0   0   1   5  15
//
// cohomology of cotangential bundle on P^3:
//
ring R=0,(x,y,z,u),dp;
resolution T1=mres(maxideal(1),0);
module M=T1[3];
intvec v=2,2,2,2,2,2;
attrib(M,"isHomog",v);
def B=sheafCoh(M,-6,2);
↦           -6  -5  -4  -3  -2  -1   0   1   2
↦        ----------------------------------------
↦     3:  70  36  15   4   -   -   -   -   -
↦     2:   -   -   -   -   -   -   -   -   -
↦     1:   -   -   -   -   -   -   1   -   -
↦     0:   -   -   -   -   -   -   -   -   6
↦        ----------------------------------------
↦  chi: -70 -36 -15  -4   0   0  -1   0   6
```

7.1.2 Exterior algebra method for sheaf cohomology computation

Let V be a $\Bbbk$-vector space of dimension $n+1$ over $\Bbbk$ and a basis $y_0, \ldots, y_n$. Let $W := V^*$ be a dual space to V with a dual basis $x_0, \ldots, x_n$. Denote the projective space of 1-quotients of W (i.e. of lines through origin in V) by $\mathbb{P}^n := \mathbb{P}(W)$, $S := \mathrm{Sym}_{\Bbbk}(W)$ its homogeneous coordinate ring, i.e. isomorphic to $\Bbbk[x_0, \ldots, x_n]$, graded by putting $\deg x_i = 1$. Let $E := \bigwedge(V)$ be the exterior algebra on V, graded by putting $\deg y_j = -1$.

For a ($\mathbb{Z}$-)graded $\Bbbk$-algebra $\mathcal{A}$ we shall denote the category of ($\mathbb{Z}$-)graded finitely generated $\mathcal{A}$-modules by gr.$\mathcal{A}$-Mod, the category of and the derived category of bounded complexes of finitely generated graded $\mathcal{A}$-modules by $\mathrm{D}^{\mathrm{b}}(\mathrm{gr}.\mathcal{A}\text{-Mod})$.

Theorem 7.1.10 (The BGG correspondence (cf. [15] and Corollary 2.7 from [36])). *The derived category of bounded complexes of finitely generated graded S-modules and the derived category of bounded complexes of finitely generated graded E-modules are equivalent.*

Remark 7.1.11 ([106, 75]). The BGG correspondence is a particular case of *Koszul duality*, which says that: the derived categories $\mathrm{D}^{\mathrm{b}}(\mathrm{gr}.\mathcal{A}\text{-Mod})$ and $\mathrm{D}^{\mathrm{b}}(\mathrm{gr}.\mathcal{A}^!\text{-Mod})$ are equivalent, where $\mathcal{A}^!$ is the Koszul dual of $\mathcal{A}$, provided $\mathcal{A}$ is Gornstein. Note moreover that $(\mathcal{A}^{n|m})^! =$

$\mathcal{A}^{m|n}$ (cf. Remark 3.1.2) and in particular, $S^! = E, E^! = S$, that is, we may think about E as $\operatorname{Ext}_S^\bullet(\Bbbk, \Bbbk)$ and about S as $\operatorname{Ext}_E^\bullet(\Bbbk, \Bbbk)$.

The BGG correspondence consists of the pair of adjoined functors $\mathbf{R}$ and $\mathbf{L}$ but for computing sheaf cohomology we are mainly interested in the following explicit construction for $\mathbf{R}(M)$, where $\mathbf{R}$ is to be considered as a functor from gr.S-Mod to the category of bounded complexes, by regarding $M \in$ gr.S-Mod as a (trivial) complex and forgetting about classes of complexes due to the construction of derived categories.

Let $M \in$ gr.S-Mod, considered as a complex concentrated in degree 0. Put

$$F^i := \operatorname{Hom}_\Bbbk(E, M_i) = M_i \otimes_\Bbbk \omega_E,$$

where M_i is considered here as a $\Bbbk$-vector space concentrated in degree i and

$$\omega_E := \operatorname{Hom}_\Bbbk(E, \Bbbk) = E \otimes \wedge^{n+1} W \cong E(-n-1).$$

Then the image $\mathbf{R}(M) \in \mathrm{D}^b(\mathrm{gr}.E\text{-Mod})$ is the following complex of E-modules:

$$\mathbf{R}(M): \cdots \leftarrow F^{i+1} \xleftarrow{\phi_i} F^i \xleftarrow{\phi_{i-1}} F^{i-1} \leftarrow \cdots,$$

with the maps

$$\phi_i : F^i \ni \alpha \mapsto \left(e \mapsto \sum_j x_j \alpha(y_j \wedge e) \right) \in F^{i+1}.$$

Proposition 7.1.12 (Propositions 2.1 and 2.3 from [36]). *The functor $\mathbf{R}$ is an equivalence between gr.S-Mod and the category of linear free complexes over E (those of which the d-th free module has socle of degree d).*

Moreover, if M is a graded S-module then $\mathrm{H}^j(\mathbf{R}(M))_i = \operatorname{Tor}^S_{i-j}(\Bbbk, M)_i$.

It is important that the complex $\mathbf{R}(M)$ is eventually exact:

Theorem 7.1.13 (Corollary 2.4 [36]). *Let $M \in$ gr.S-Mod and let r be its Castelnuovo-Mumford regularity[1], then the complex $\mathbf{R}(M)$ is exact at F^i for all $i \geqslant s$ iff $s > r$.*

Moreover the truncated complex $\mathbf{R}(M)_{>r}$ is exact and all remaining maps are linear:

$$\mathbf{R}(M)_{>r} : \cdots \leftarrow F^{r+3} \xleftarrow{\phi_{r+2}} F^{r+2} \xleftarrow{\phi_{r+1}} F^{r+1}.$$

Definition 7.1.14. The **Tate resolution** of M is the following doubly infinite free exact complex $\mathbf{T}(M)$ of E-modules:

$$\mathbf{T}(M) : \cdots \leftarrow F^{r+3} \xleftarrow{\phi_{r+2}} F^{r+2} \xleftarrow{\phi_{r+1}} F^{r+1} \leftarrow \mathbf{L}^r \leftarrow \mathbf{L}^{r-1} \leftarrow \cdots,$$

---
[1]If M is a finitely generated S-module than for all $r \gg 0$: $M_{\geqslant r}$ is generated in degree r and has a linear free resolution. The **Castelnuovo-Mumford regularity** of M is the least such integer r (cf. [34, Chapter 20])

7.1. PROJECTIVE GEOMETRY

which is constructed by adjoining a minimal free resolution (over E) $\mathbf{L}^\bullet$ of $\mathrm{Ker}\left(F^{r+2} \xleftarrow{\phi_{r+1}} F^{r+1}\right)$:

$$\mathbf{L}_\bullet : 0 \leftarrow \mathrm{Ker}(\phi_{r+1}) \leftarrow \mathbf{L}^r \leftarrow \mathbf{L}^{r-1} \leftarrow \cdots$$

to the right of $\mathbf{R}(M)_{>r}$.

In fact, we can construct $\mathbf{T}(M)$ by starting from any truncation $\mathbf{R}(M_{>s})$, $s \geq r$. Therefore we have the following:

Remark 7.1.15. The Tate resolution $\mathbf{T}(M)$ only depends on the sheaf $\mathcal{F} = \widetilde{M}$. The complex $\mathbf{T}(\mathcal{F}) := \mathbf{T}(M)$ is called the **Tate resolution** of $\mathcal{F}$.

Examples 7.1.16 ([29]).

If M has finite length then $\mathbf{T}(M) : \cdots \leftarrow 0 \leftarrow 0 \leftarrow 0 \leftarrow \cdots$,

If $M = S$ then $\mathbf{T}(M) : \cdots \leftarrow W \otimes_{\Bbbk} \omega_E \leftarrow \omega_E \leftarrow E \leftarrow W^* \otimes_{\Bbbk} E \leftarrow \cdots$.

Theorem 7.1.17 (Theorem 4.1 and Corollary 4.2 from [36]). *If $\mathcal{F} \in \mathrm{Coh}\mathbf{P}(W)$ then the linear part of the Tate resolution $\mathbf{T}(\mathcal{F})$ is $\bigoplus_j \mathbf{R}(\bigoplus_d \mathrm{H}^j\mathcal{F}(d))$. In particular, the term of the complex $\mathbf{T}(\mathcal{F})$ with cohomological degree i is*

$$\mathbf{T}(\mathcal{F})^{(i)} = \bigoplus_j \mathrm{Hom}_{\Bbbk}\left(E, \mathrm{H}^j\mathcal{F}(i-j)\right) = \bigoplus_j \mathrm{H}^j\mathcal{F}(i-j) \otimes_{\Bbbk} \omega_E,$$

where $\mathrm{H}^j\mathcal{F}(i)$ is regarded as a vector space concentrated in degree i.
That is, for any $j, i \in \mathbb{Z}$: $\mathrm{H}^j\mathcal{F}(i) = \mathrm{Hom}_E\left(\Bbbk, \mathbf{T}(\mathcal{F})^{(i+j)}\right)_{-i}$.

Observing that each cohomology group of each twist of $\mathcal{F}$ occurs once in a term of $\mathbf{T}(\mathcal{F})$, we can compute part of the cohomology of $\mathcal{F}$ by computing part of the Tate resolution.

The exterior algebra method for computing sheaf cohomology relies on computing free resolutions over E for computing the Tate resolution $\mathbf{T}^{>r}(M)$, and has been implemented in SINGULAR (cf. command `sheafCohBGG` (cf. 7.1.2) and `sheafCohBGG2` (cf. 7.1.2) from `sheafcoh.lib`). Let us give a part of User Manual (cf. [118]) describing these procedures:

sheafCohBGG

Usage:
  sheafCohBGG(M,l,h); M module, l,h int

Assume:
  M is graded, and it comes assigned with an admissible degree vector as an attribute, `h>=l`, and the basering has `n+1` variables.

112   CHAPTER 7. APPLICATIONS

Return:
   intmat, cohomology of twists of the coherent sheaf F on P^n associated to coker(M).
   The range of twists is determined by l, h.

Display:
   The intmat is displayed in a diagram of the following form:

```
                    l              l+1                        h
         -----------------------------------------------------------
      n:       h^n(F(l))      h^n(F(l+1))       ......    h^n(F(h))
              ..........................................................
      1:       h^1(F(l))      h^1(F(l+1))       ......    h^1(F(h))
      0:       h^0(F(l))      h^0(F(l+1))       ......    h^0(F(h))
         -----------------------------------------------------------
    chi:       chi(F(l))      chi(F(l+1))       ......    chi(F(h))
```

   A '-' in the diagram refers to a zero entry; a '*' refers to a negative entry (=
   dimension not yet determined). refers to a not computed dimension.

Note:
   This procedure is based on the Bernstein-Gel'fand-Gel'fand correspondence and on
   Tate resolution (see [Eisenbud, Floystad, Schreyer: Sheaf cohomology and free resolutions over exterior algebras, Trans AMS 355 (2003)]).
   sheafCohBGG(M,l,h) does not compute all values in the above table. To determine
   all values of h^i(F(d)), d=l..h, use sheafCohBGG(M,l-n,h+n).

Example:

```
LIB "sheafcoh.lib";
// cohomology of structure sheaf on P^4:
//----------------------------------------
ring r=0,x(1..5),dp;
module M=0;
def A=sheafCohBGG(M,-9,4);
↦       -9  -8  -7  -6  -5  -4  -3  -2  -1   0   1   2   3   4
↦    ----------------------------------------------------------------
↦    4:  70  35  15   5   1   -   -   -   -   *   *   *   *   *
↦    3:   *   -   -   -   -   -   -   -   -   -   *   *   *   *
↦    2:   *   *   -   -   -   -   -   -   -   -   -   -   *   *
↦    1:   *   *   *   -   -   -   -   -   -   -   -   -   -   *
↦    0:   *   *   *   *   -   -   -   -   -   1   5  15  35  70
↦    ----------------------------------------------------------------
↦  chi:   *   *   *   *   1   0   0   0   1   *   *   *   *
```

7.1. PROJECTIVE GEOMETRY 113

```
// cohomology of cotangential bundle on P^3:
//-----------------------------------------
ring R=0,(x,y,z,u),dp;
resolution T1=mres(maxideal(1),0);
module M=T1[3];
intvec v=2,2,2,2,2,2;
attrib(M,"isHomog",v);
def B=sheafCohBGG(M,-8,4);
↦          -8   -7   -6   -5   -4   -3   -2   -1    0    1    2    3    4
↦     ----------------------------------------------------------------------
↦     3:  189  120   70   36   15    4    -    -    -    -    *    *    *
↦     2:    *    -    -    -    -    -    -    -    -    -    -    *    *
↦     1:    *    *    -    -    -    -    -    -    1    -    -    -    *
↦     0:    *    *    *    -    -    -    -    -    -    -    6   20   45
↦     ----------------------------------------------------------------------
↦   chi:    *    *    *  -36  -15   -4    0    0   -1    0    *    *    *
```

sheafCohBGG2

Usage:
    sheafCohBGG2(M,l,h); M module, l,h int

Assume:
    M is graded, and it comes assigned with an admissible degree vector as an attribute, h>=l, and the basering has n+1 variables.

Return:
    intmat, cohomology of twists of the coherent sheaf F on P^n associated to coker(M). The range of twists is determined by l, h.

Display:
    The intmat is displayed in a diagram of the following form:

```
                       l              l+1                       h
           -----------------------------------------------------------
      n:         h^n(F(l))      h^n(F(l+1))      ......    h^n(F(h))
           ...............................................
      1:         h^1(F(l))      h^1(F(l+1))      ......    h^1(F(h))
      0:         h^0(F(l))      h^0(F(l+1))      ......    h^0(F(h))
           -----------------------------------------------------------
    chi:         chi(F(l))      chi(F(l+1))      ......    chi(F(h))
```

A '-' in the diagram refers to a zero entry; a '*' refers to a negative entry (= dimension not yet determined). refers to a not computed dimension.

CHAPTER 7. APPLICATIONS

If `printlevel>=1`, step-by step timings will be printed. If `printlevel>=2` we add progress debug messages if `printlevel>=3`, even all intermediate results...

Note:
This procedure is based on the Bernstein-Gel'fand-Gel'fand correspondence and on Tate resolution (see [Eisenbud, Floystad, Schreyer: Sheaf cohomology and free resolutions over exterior algebras, Trans AMS 355 (2003)]).
`sheafCohBGG(M,l,h)` does not compute all values in the above table. To determine all values of `h^i(F(d))`, d=l..h, use `sheafCohBGG2(M,l-n,h+n)`. Experimental version: require less memory and uses speedups due to Dissertation by O. Motsak (2010).

Example:

```
LIB "sheafcoh.lib";
int pl = printlevel;
int l = -6; int h = 6; // range of twists: l..h
//-----------------------------------------------
// Abelian surface in P^4, deg: 10, gen: 6 (B6.1)
//-----------------------------------------------
ring R = (31991),(x,y,z,u,v),dp;
ideal I=-x^3*y^2*z-7318*x^4*z^2-x^2*z^3*u-x*y^3*u^2-14636*x^2*y*z*u^2-y*z^2*u^3
-7318*y^2*u^4+y^5*v-8856*x^2*y*z^2*v+z^5*v-8856*y^2*z*u^2*v-7318*x^3*y*v^2
+5535*x*z^3*v^2+5535*y^3*u*v^2-4*x*y*z*u*v^2-7318*z*u^3*v^2-8856*x*y^2*v^3
-8856*z^2*u*v^3+5535*y*z*v^4+7318*x*u*v^4+v^6,-x*y^2*z*u^2-7318*x^2*z^2*u^2
-z^3*u^3-7318*y*z*u^4+y^4*z*v+7318*x*y^2*z^2*v-7318*z^4*u*v-8856*y*z^2*u^2*v
+x^2*y^2*v^2+5535*y^2*z*u*v^2-x*z^2*u*v^2+7318*x*y*u^2*v^2+7318*y^3*v^3+y*u*v^4,
-x*y^2*u^3-7318*x^2*z*u^3-z^2*u^4-7318*y*u^5+y^4*u*v+7318*x*y^2*z*u*v
-7318*z^3*u^2*v-8856*y*z*u^3*v-7318*x^3*v^2+5535*y^2*u^2*v^2-2*x*z*u^2*v^2
-y^2*z*v^3-8856*x*z^2*v^3-8856*x*y*u*v^3-x^2*v^4-7318*y*v^5,y^5*z+7318*x*y^3*z^2
-7318*y*z^4+x^3*y*u^2-8856*y^2*z^2*u^2-5535*x*y*v+14636*x*y^2*u^2*v+7318*z^2*u^3
+7318*y^4*v^2+x*u^3*v^2+y^2*u*v^3,y^4*z^2+7318*x*y^2*z^3-7318*z^5+u+x^3*z*u^2
-8856*y*z^3*u^2+5535*x*z^2*u^3-x*y*u^4-7318*u^6+2*x^2*y^2*z*v+7318*x^3*z^2*v
+5535*y^2*z^2*u*v+y^3*u^2*v-2719*x*y*z*u^2*v-8856*x*u^4+x^4*v^2+7318*y^3*z*v^2
+5535*x^2*z*u*v^2+5535*y*u^3*v^2+7318*x^2*y*v^3-8856*x*u^2*v^3-7318*u*v^5,
-x*y^2*z^3-7318*x^2*z^4-z^5*u-7318*y*z^3*u^2-x^3*z^2-5535*x*z*x^2*z*u+x*y*z*u^2*v
+7318*z*u^4-y^3*z*v^2-14636*x*y*z^2*u^2+8856*z^2*u^2*v^2-x^2*v^3
-5535*y*z*u*v^3-7318*x*u^2*v^3-7318*y^2*v^4-u*v^5,y^3*z^3+7318*x*y*z^4-x^4*y*u
+7318*y^4*z*u+8856*x*y^2*z^2*u-5535*x^2*y*z*u^2-x^3*u^3-x*z*u^4+x^2*y*z^2*v
-7318*x^2*y^2*u*v+y^2*z*u^2*v-7318*x*z*u^2*v+7318*y^2*z^2*v-x^2*u^2*v^2,
-y^2*z^4-7318*x*z^5+x^4*v-4*x*y^2*z*u-8856*x*y*z^2*u-x*y^2*v-8856*y*z^2*u-5535*x*z^2*u^2
-x^2*y*u^3-7318*x*u^5-x^2*z^3*v+7318*x^2*y*z*u*v-2*y*z^2*u*v-7318*y^2*u^3*v-7318*y^3*u^3*v
-8856*x*x*z*u^3*v-7318*y*z^3*v^2-u^4*v^2-7318*z*u^2*v^3,y^6+7318*x*y^4*z
-7318*y^2*z^3*u-7318*x^4*u^2-8856*y^3*z*u^2+5535*x*y^2*u^3-x^2*z^2*u^3+y*u^5
-7318*x^3*y^2*v+5535*y^4*u*v-4*x*y^2*z*u*v-14636*x^2*z^2*u*v-8856*x^2*y*u^2*v
-z^3*u^2*v-8856*x*y^3*v^2-7318*z^2*u^2*v^2-x^3*u*v^2+5535*y^2*z*v^3
-x*x*z^2*v+y*v^5,x^4*u+7318*x^3*y*z-8856*x*y^3*v^2-7318*z^2*u^3+5535*x*z^2*u^2-7318*x*u^5+7318*x^2*y^3*v
+5535*y^2*z^3*v-x*z^4*v-2719*x*y*z^2*u*v-8856*z^2*u^3*v+v^2*x^2*y*u*v^2
```

7.1. PROJECTIVE GEOMETRY

```
+5535*y*z*u^2*v^2+7318*x*u^3*v^2+y*z^2*v^3+7318*y^2*u*v^3+u^2*v^4,-x^3*y^3
-7318*x^4*y*z-x^2*y*z^2*u-7318*x^2*y^2*u^2-7318*x*y^4*v-8856*x^2*y^2*z*v+y*z^4*v
+7318*y^2*z^2*u*v+5535*x*y*z^2*v^2-x*y^2*u*v^2+7318*x^2*z*u*v^2+z^2*u^2*v^2
+7318*z^3*v^3+x*z*v^4,-x^2*y^4-7318*x^3*y^2*z-2*x*y^2*z^2*u-7318*x^2*z^3*u
-7318*x*y^3*u^2-z^4*u^2-7318*y*z^2*u^3-7318*y^5*v-8856*x*y^3*z*v-7318*z^5*v
-8856*y*z^3*u*v+5535*y^2*z^2*v^2-x*z^3*v^2-y^3*u*v^2+7318*x*y*z*u*v^2+y*z*v^4,
-x*y^5-7318*x^2*y^3*z-y^3*z^2*u-7318*y^4*u^2+7318*x^4*y*v-5535*x*y^3*u*v
+x^2*y*z*u*v-y^2*u^3*v+8856*x^2*y^2*v^2-y*z^3*v^2-14636*y^2*z*u*v^2
-5535*x*y*z*v^3-7318*x^2*u*v^3-z*u^2*v^3-7318*z^2*v^4-x*v^5,-x*y^4*u
-7318*x^2*y^2*z*u-y^2*z^2*u^2-7318*y^3*u^3+7318*x^4*u*v-5535*x*y^2*u^2*v
+x^2*z*u^2*v-y*u^4*v+x*y^2*z*v^2+7318*x^2*z^2*v^2+8856*x^2*y*u*v^2
-7318*y*z*u^2*v^2+x^3*v^3+7318*x*y*v^4,-x^2*y^2*z^2-7318*x^3*z^3-x*z^4*u
-7318*x*y*z^2*u^2-x^4*z*v-5535*x^2*z^2*u*v+x^2*y*u^2*v+7318*x*u^4*v
-7318*x^2*y*z*v^2+y*z^2*u*v^2+7318*y^2*u^2*v^2+8856*x*z*u^2*v^2+u^3*v^3
+7318*z*u*v^4,-x*y^3*z-7318*x^2*y*z^2-y*z^3*u-7318*y^2*z*u^2-x^3*y*v
-5535*x*y*z*u*v-7318*x^2*u^2*v-z*u^3*v-7318*x*y^2*v^2-7318*z^2*u*v^2-x*u*v^3,
7318*x^5+7318*y^5+8856*x*y^3*z+7318*z^5-5535*x^2*y^2*u+8856*y*z^3*u
-5535*x*z^2*u^2+7318*u^5+8856*x^3*y*v-5535*y^2*z^2*v-4599*x*y*z*u*v+8856*z*u^3*v
-5535*x^2*z*v^2-5535*y*u^2*v^2+8856*x*u*v^3+7318*v^5,-x^2*y^2*u-7318*x^3*z*u
-x*z^2*u^2-7318*x*y*u^3-y^2*z^2*v-7318*x*z^3*v+x*z^2*v^2+7318*y^3*u*v-8856*x*y*z*u*v
-x^2*z*v^2-y*u^2*v^2-7318*y*z*v^3,-x^5-y^5+8856*x^2*y*z^2-z^5-5535*x^3*z*u
+8856*y^2*z*u^2-5535*x*y*u^3-u^5-5535*x*z^3*u*v-5535*y^3*u*v+5*x*y*z*u*v
+8856*x^2*u^2*v+8856*x*y^2*v^2+8856*z^2*u*v^2-5535*y*z*v^3-v^5;
resolution FI = mres(I,2); module M=FI[2];
////////////////////////////////////////////
printlevel = 0;
int t = timer;
def B = sheafCoh(M, 1, h); // global Ext method:
```

	-6	-5	-4	-3	-2	-1	0	1	2	3	4	5	6
4:	5	1	-	-	-	-	-	-	-	-	-	-	-
3:	180	125	80	45	20	5	1	-	-	-	-	-	-
2:	-	-	-	-	-	-	2	-	-	-	-	-	-
1:	-	-	-	-	-	-	-	-	5	10	10	2	-
0:	-	-	-	-	-	-	-	-	-	-	-	3	30
chi:	-175	-124	-80	-45	-20	-5	1	0	-5	-10	-10	1	30

```
"Time: ", timer - t;
↦ Time:  1544
////////////////////////////////////////////
t = timer;
B = sheafCohBGG(M, 1, h); // BGG method (without optimization):
```

	-6	-5	-4	-3	-2	-1	0	1	2	3	4	5	6
4:	5	1	-	-	-	-	-	-	-	-	-	-	-
3:	*	125	80	45	20	5	1	-	-	-	-	-	-
2:	*	*	-	-	-	-	2	-	-	-	-	-	-
1:	*	*	*	-	-	-	-	-	5	10	10	2	-
0:	*	*	*	*	-	-	-	-	-	-	-	3	30
chi:	*	*	*	*	-20	-5	1	0	-5	-10	-10	1	*

```
"Time: ", timer - t;
↦ Time: 10
//////////////////////////////////////////
t = timer;
B = sheafCohBGG2(M, l, h); // BGG method (with optimization)
↦ Cohomology table:
↦        -6   -5  -4  -3  -2  -1   0   1   2   3   4   5   6
↦       ---------------------------------------------------------------
↦   4:   5    1   -   -   -   -   -   -   -   -   -   -   -
↦   3:   *   125  80  45  20   5   1   -   -   -   -   -   -
↦   2:   *    *   -   -   -   -   2   -   -   -   -   -   -
↦   1:   *    *   *   -   -   -   -   -   5  10  10   2   -
↦   0:   *    *   *   *   -   -   -   -   -   -   -   3  30
↦       ---------------------------------------------------------------
↦ chi:   *    *   *   *  -20  -5   1   0  -5 -10 -10   1   *
"Time: ", timer - t;
↦ Time: 4
//////////////////////////////////////////
printlevel = pl;
```

7.1.3 Sheaf cohomology: benchmarks

Starting from SINGULAR $3-0-0$, computation of Tate resolutions in `sheafCohBGG` and `sheafCohBGG2` is done by Algorithm 5.1.2 using the improvements to the GB-Algorithm described in this thesis.

Moreover, the command `sheafCohBGG2` (cf. 7.1.2) also incorporates all our performance tweaks for truncation and other time consuming computations over a commutative polynomial algebra.

In order to test/benchmark our improvements we computed sheaf cohomology tables for given ranges of twists (from $-7..7$ up to $-12..12$) of the surfaces listed in [28]. The defining ideals were computed using the Macaulay Classic code provided by W. Decker. Each surface was tested twice. №2 uses as input the original ideal due to Macaulay Classic. №1 uses as input a GB (w.r.t. `dp`) of the ideal used in №2.

We compared 64-bit SINGULAR 3.0 (denoted by SCA) and M2 v. 1.2 (generic 64-bit built with `libc-2.7`). Computations were performed on a Gentoo Linux (2.6.30) server with *AMD Opteron(tm) Processor* 242×2 running at (1590.741 Mhz) and 8 Gb memory. Timings on tables 7.1 and 7.2 are given in $\frac{1}{100}$-th of a second. Afterwards we illustrate the good asymptotic behavior of SCA (dark gray plot) compared to M2(light gray plot) by plotting the timings (logarithmically in seconds) depending on the number of twists (where $-r..r$ corresponds to $2r$). Captions name corresponding example from top to bottom and from left to right. More details can be found online at http://www.mathematik.uni-kl.de/~motsak/tests/.

Judging from our benchmarks, M2 is more sensitive to the input data (depends on the

7.1. PROJECTIVE GEOMETRY

defining ideal), and performs better on smaller examples, while SINGULAR seems to be input insensitive and has a better asymptotic performance (that is, much better on harder examples)

While the computation of the resolution over the exterior algebra is the major time-consuming part of the computation, in some cases the computation of Castelnuovo-Mumford (CM)regularity took the leadership.

We excluded the example k3.d10.g9.quart2 from the tables since M2 needed several days just to compute CM-regularity.

Surface	№	-7..7		-8..8		-9..9		-10..10	
		M2	SCA	M2	SCA	M2	SCA	M2	SCA
bielliptic.d10.g6	1	130	115	142	142	693	502	2802	1986
	2	133	116	143	140	731	521	3150	1996
bordiga	1	109	90	556	348	2535	1203	7760	3897
	2	117	89	582	353	2600	1203	7942	4000
castelnuovo	1	104	86	427	280	1684	852	5226	2269
	2	145	88	1103	288	5319	839	19710	2344
cubicscroll	1	79	81	265	259	855	748	2309	2055
	2	89	81	280	262	878	772	2417	2055
ell.d10.g10	1	538	506	2877	1753	13289	6024	56322	21039
	2	528	507	2831	1698	13569	6055	56537	21042
ell.d10.g9	1	3270	1987	4900	2462	19035	10250	74683	43656
	2	3576	2077	5164	2565	21144	10257	82254	41192
ell.d11.g12	1	1333	836	5996	2686	22020	10132	93886	35778
	2	1424	840	6410	2581	24264	9700	99489	38040
ell.d12.g13	1	2785	1717	3487	2560	13301	8717	54574	27241
	2	2796	1641	3484	2579	13376	8638	54712	30371
ell.d12.g14.ss0	1	4195	1540	9661	2569	28718	7752	91448	25988
	2	4226	1527	9710	2565	28975	7440	92416	26556
ell.d12.g14.ssue	1	3690	1610	7169	2717	25057	8510	95381	28593
	2	3100	1611	6536	2829	21172	8336	81177	27901

ell.d8.g7	1	101	243	523	599	1813	1474	5396	3367
	2	102	286	533	745	1841	2264	5473	5730
ell.d9.g7	1	971	555	4793	2361	19489	7779	73166	23641
	2	1024	556	5022	2455	20062	7792	78401	22242
elliptic.scroll	1	580	240	3044	1115	12140	3557	40805	13347
	2	569	242	3016	1116	11981	3724	40419	13364
enr.d10.g8	1	2266	1278	2836	3527	15468	17232	71424	69402
	2	1327	1241	1873	3489	11380	17333	53234	69207
enr.d11.g10	1	4402	2047	5238	2413	18137	9219	76750	28802
	2	3555	2037	4410	2543	18036	9928	81976	28172
enr.d13.g16	1	1523	735	2958	1176	9557	3904	35982	15435
	2	1519	735	2960	1201	9541	3706	35471	11316
enr.d13.g16.two	1	1825	783	3246	1192	11085	6080	41422	26910
	2	1812	767	3261	1229	11142	4643	41245	13744
enr.d9.g6	1	2753	1770	2892	1736	13753	7682	55844	24616
	2	2827	1774	2960	1743	13878	7682	55599	24701
k3.d10.g9.quart1	1	3375	2229	5029	2693	18732	11836	70357	48826
	2	3664	2328	5373	2813	21535	12507	85071	48631
k3.d11.g11.ss0	1	1879	1300	2977	1983	11192	7109	41789	23261
	2	1882	1294	2959	1987	11382	7135	41964	24536
k3.d11.g11.ss1	1	2682	1699	3768	2436	16023	9346	67633	32445
	2	2731	1558	3808	2383	16551	8440	68836	33785
k3.d11.g11.ss2	1	2881	1261	3957	1958	17756	6570	75557	20949
	2	2444	1281	3524	2041	16677	6124	71101	17158
k3.d11.g11.ss3	1	3600	1884	4709	2649	16716	6029	62943	16167
	2	2648	1832	3564	2563	11941	5876	48776	15438
k3.d11.g12	1	950	758	3272	1997	12248	4933	50183	15971
	2	938	759	3273	1935	12274	4950	49221	15273

7.1. PROJECTIVE GEOMETRY

k3.d12.g14	1	1160	1014	2980	2510	8963	7614	29529	23519
	2	1148	1018	2997	2518	9004	7662	29717	23535
k3.d13.g16	1	8554	3996	9756	5693	17177	11450	46059	22918
	2	8831	4268	10082	5913	17672	11706	46515	22539
k3.d14.g19	1	3681	3448	5956	6338	13389	12979	36913	26709
	2	3745	3652	5972	6157	13436	13195	37894	27398
k3.d7.g5	1	61	186	299	591	1801	1863	7373	5432
	2	62	183	299	592	1804	1919	7352	5451
k3.d8.g6	1	441	384	2626	1899	11550	7276	43705	24967
	2	467	383	3014	1899	14367	7054	53288	24280
k3.d9.g8	1	295	317	1617	1179	7491	4008	30033	12987
	2	293	321	1637	1215	7560	4020	30260	13043
rat.d10.g8	1	42803	13562	44493	13205	59220	23717	119729	45103
	2	2151	13523	3195	13212	12677	21542	48786	44640
rat.d10.g9.quart1	1	800	245	3881	1710	16400	9023	62317	40478
	2	949	254	4608	1713	20603	9097	79992	37995
rat.d10.g9.quart2	1	2746	1029	3545	1062	12132	4809	55141	22415
	2	2808	994	3601	1103	12378	4621	52226	21210
rat.d11.g11.ss0	1	1992	1062	3090	1234	11291	5657	41348	22342
	2	2036	1029	3140	1227	11410	5658	41907	21116
rat.d11.g11.ss1	1	1892	934	2956	1106	11984	5043	48181	26342
	2	1700	909	2797	1107	11019	5283	47188	25153
rat.d11.g11.ssue	1	3202	1128	3945	1332	19451	6880	83916	29266
	2	2472	1089	3215	1291	15267	6958	66801	28416
rat.d7.g4	1	587	329	3038	1847	14162	8629	57279	32679
	2	965	330	5899	1931	27531	9084	114296	30627
rat.d8.g5	1	496	284	4341	1986	16777	8716	60314	29485
	2	35037	276	40253	1913	53988	8324	114703	27662

Surface	№									
rat.d8.g6	1	347	241	1910	1228	10501	5451	44911	21129
	2	371	243	2005	1230	10756	5436	50136	21201
rat.d9.g6	1	4928	2652	5076	2818	25114	12705	100531	42592
	2	3817	2781	3950	2823	20711	12755	86637	42707
rat.d9.g7	1	1048	515	4596	1703	14114	6110	60867	23239
	2	979	505	4242	1698	14912	6111	64330	22472
veronese	1	1336	453	5847	1595	21602	5379	68617	14039
	2	1077	439	5014	1604	18244	5654	58668	13991

Table 7.1: Benchmarks for sheaf cohomology computations with M2 and SINGULAR with a clear advantage of SINGULAR

Surface	№	-7..7		-8..8		-9..9		-10..10		-12..12	
		M2	SCA	M2	SCA	M2	SCA	M2	SCA	M2	SCA
ab.d10.g6	1	496	335	498	368	504	432	2476	1224	33543	11561
	2	140	333	144	382	151	448	890	1278	17135	11783
ab.d15.g21	1	1166	774	1228	844	1220	1029	1412	1653	8095	7446
	2	945	762	985	862	1001	1040	1143	1649	5576	7443
ab.d15.g21.quint1	1	86	157	155	289	389	597	1197	1390	15753	11507
	2	81	156	150	296	357	611	1039	1437	14493	11535
bielliptic.d15.g21	1	96	142	172	262	400	579	1261	1474	14340	12623
	2	93	143	172	268	390	575	1080	1511	10063	12647
ell.d7.g6	1	57	218	146	512	454	1248	1752	2761	19702	15444
	2	57	217	147	514	608	1277	4601	2687	70573	14770

Table 7.2: Benchmarks for sheaf cohomology computations with M2 and SINGULAR

7.1. PROJECTIVE GEOMETRY

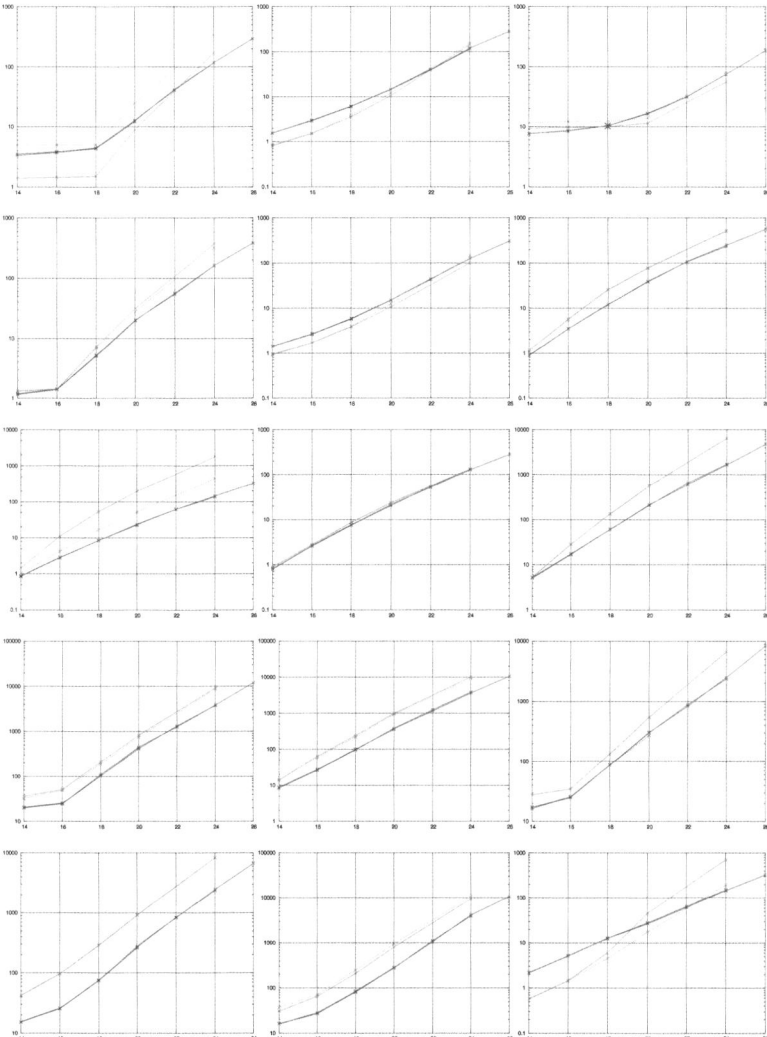

Figure 7.1: ab.d10.g6, ab.d15.g21.quint1, ab.d15.g21; bielliptic.d10.g6, bielliptic.d15.g21, bordiga; castelnuovo, cubicscroll, ell.d10.g10; ell.d10.g9, ell.d11.g12, ell.d12.g13; ell.d12.g14.ss0, ell.d12.g14.ssue, ell.d7.g6;

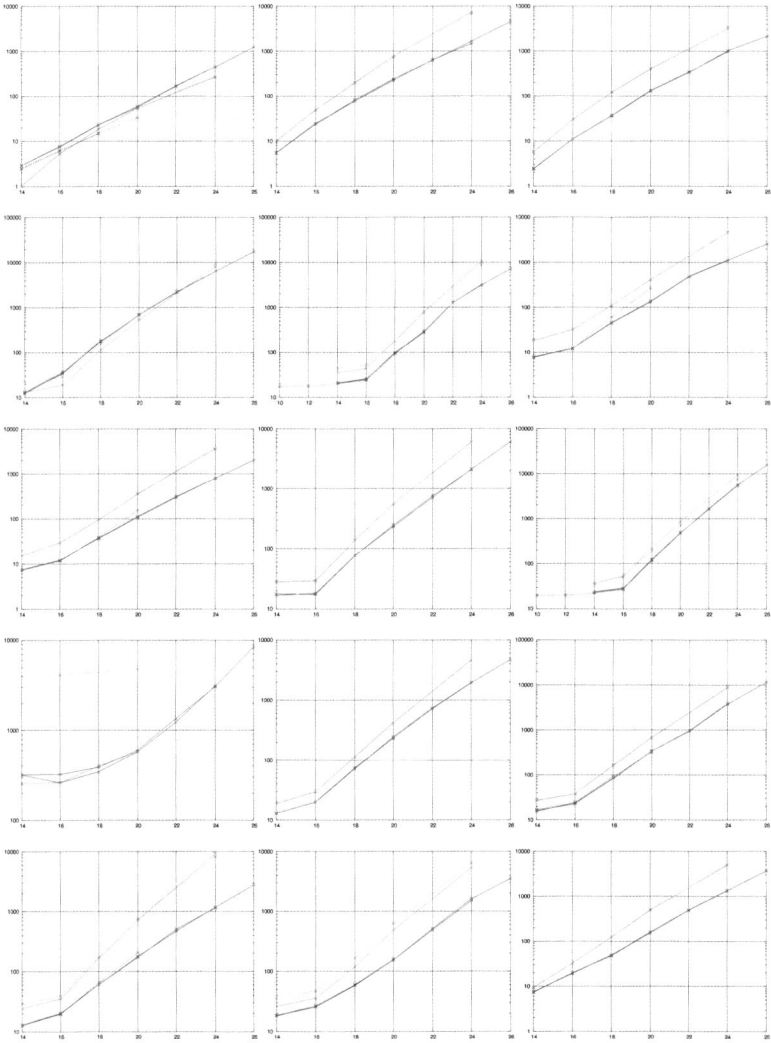

Figure 7.2: ell.d8.g7, ell.d9.g7, elliptic.scroll; enr.d10.g8, enr.d11.g10, enr.d13.g16.two; enr.d13.g16, enr.d9.g6, k3.d10.g9.quart1; k3.d10.g9.quart2, k3.d11.g11.ss0, k3.d11.g11.ss1; k3.d11.g11.ss2, k3.d11.g11.ss3, k3.d11.g12;

7.1. PROJECTIVE GEOMETRY

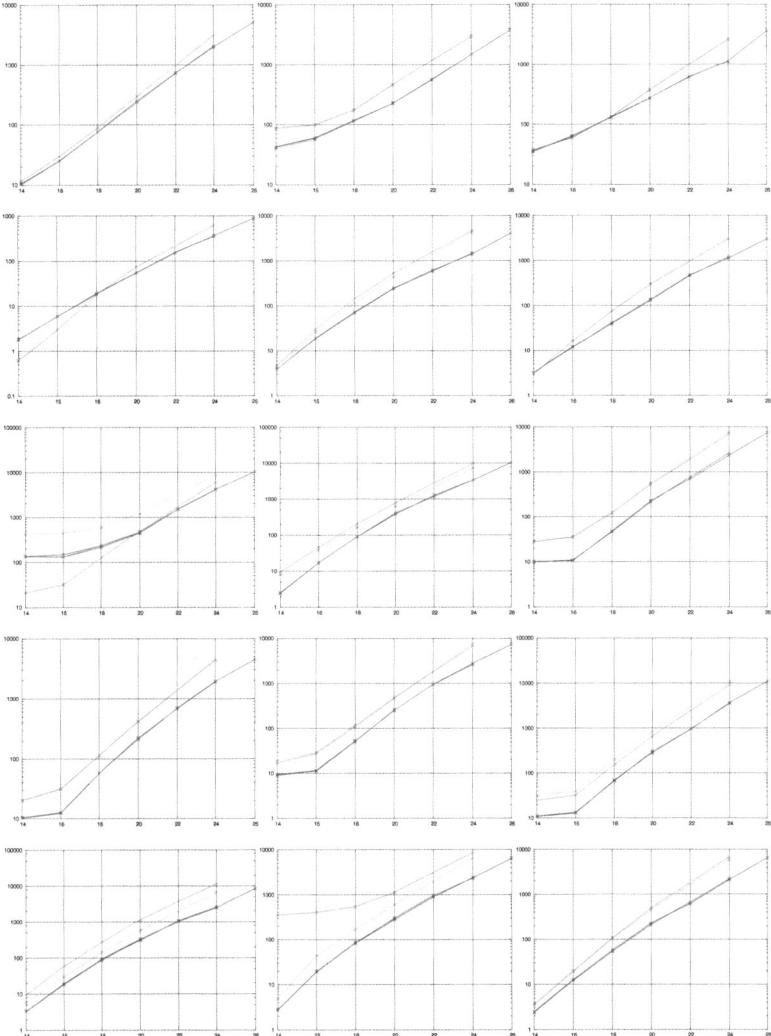

Figure 7.3: k3.d12.g14, k3.d13.g16, k3.d14.g19; k3.d7.g5, k3.d8.g6, k3.d9.g8; rat.d10.g8, rat.d10.g9.quart1, rat.d10.g9.quart2; rat.d11.g11.ss0, rat.d11.g11.ss1, rat.d11.g11.ssue; rat.d7.g4, rat.d8.g5, rat.d8.g6;

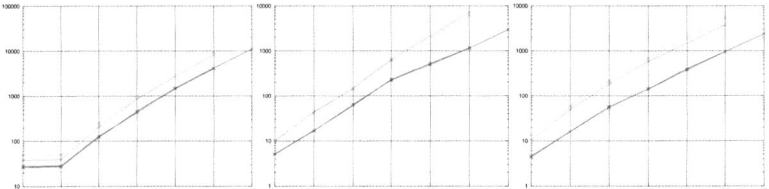

Figure 7.4: rat.d9.g6, rat.d9.g7, veronese;

7.2 Coordinate-free verification of affine geometry theorems

There are several coordinate-free approaches to automated reasoning in Affine Geometry: area method, bracket algebra method, Clifford algebra methods (cf. [22] and references thereof). The main advantage of such approaches is the ability to provide geometric interpretation for all intermediate identities.

In this section we try to explain why the exterior algebra may be considered as the natural domain in which to state and prove theorems in linear and affine geometry.

There are two fundamentally different approaches to affine geometry by means of an exterior algebra of a linear space. One is due to H. Grassmann's work: "Die Ausdehnungslehre" [60] (which is the origin of *multilinear algebra*). who considered exterior algebra to be generated by points of linear space as the ring generated by numbers and points, with the only condition being the anti-commutativity of *points multiplication*. This way much of Affine Geometry reduces to a verification of identities. For example the identities (in the exterior algebra generated by points A, B, C, D)

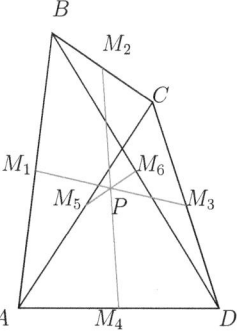

Figure 7.5: Tetrahedron

$$\frac{1}{2}\left(\frac{1}{2}(B+C)+\frac{1}{2}(D+A)\right) = \frac{1}{2}\left(\frac{1}{2}(A+B)+\frac{1}{2}(C+D)\right),$$
$$\frac{1}{2}\left(\frac{1}{2}(B+C)+\frac{1}{2}(D+A)\right) = \frac{1}{2}\left(\frac{1}{2}(B+D)+\frac{1}{2}(C+A)\right)$$

imply that the three line segments joining mid-points of opposite edges of a tetrahedron intersect at their mid-points (cf. Figure 7.5).

And the usual (modern) geometric interpretation, where elements are considered as (real) vectors from the origin (i.e. implicitly $\mathbb{R}^d$, for some implicitly fixed dimension $d \in \mathbb{N}$), and the product $A \wedge B$ of two "points" is again a "point" from origin, defined similarly to the cross product $[A, B]$ for $A, B \in \mathbb{R}^d$.

Later came ideas about using Computer Algebra over exterior and Clifford algebras for coordinate-free verification of affine geometry theorems reformulated by means of identities in exterior algebras,

D. Fearnley-Sander followed Grassmann's interpretation (cf. [49, 48, 51, 50]) while others used the other interpretation (e.g. D. Wang in [128, 127], and recently this approach was adopted/reinvented by I. Tchoupaeva in [23, 124]).

CHAPTER 7. APPLICATIONS

Either way affine geometry theorems can be formally reformulated as (homogeneous) ideal membership problems.

Using the second approach one can express the following statements:

1. Three "points" A, B and C are *collinear* iff
$$(B - A) \wedge (C - A) = 0.$$

2. Two lines, one via "points" A_1, A_2 and the other via B_1, B_2, are *parallel* iff
$$(A_2 - A_1) \wedge (B_2 - B_1) = 0.$$

3. A "point" M **divides the interval** $[A; B]$ **with the ratio** $a : b$ division of interval iff
$$b(A - M) - a(B - M) = 0.$$

Unfortunately, only statements about parallel and intersecting subspaces of $\mathbb{R}^d$ can be written in terms of outer product of vectors. For instance, this excludes statements about angles and circles.

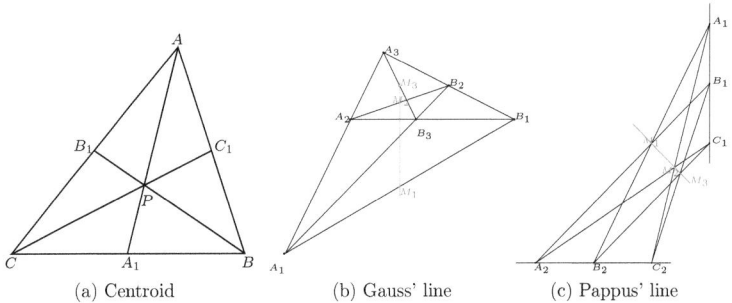

(a) Centroid   (b) Gauss' line   (c) Pappus' line

Below we will illustrate (partly following [128]) the use of SINGULAR on the following geometrical statements:

Centroid Let ABC be any triangle and A_1, B_1, C_1 be the midpoints of the three sides BC, CA, AB. Then the three lines $AA_1, BB_1 CC_1$ are concurrent (cf. Figure 7.6a).

Gauss' line theorem Let A_1, A_2, B_1, B_2 be any fixed points. Assume that A_1A_2 and B_1B_2 intersect (denote the intersection by A_3 and A_1B_2 and B_1A_2 intersect (denote the intersection by B_3. Now let M_1 be the middle of A_1B_1, M_2 be the middle of A_2B_2 and M_3 be the middle of A_3B_3.
Then the points M_1, M_2, M_3 are collinear (cf. Figure 7.6b).

7.2. COORDINATE-FREE VERIFICATION OF AFFINE GEOMETRY THEOREMS

Pappus' line theorem Let A_1, A_2, B_1, B_2 be any fixed points, C_i be point collinar with A_i and $B_i, i = 1, 2$.
Assume that A_1B_2 and A_2B_1 intersect (denote the intersection by M_1, A_1C_2 and A_2C_1 intersect (denote the intersection by M_2, B_1C_2 and B_2C_1 intersect (denote the intersection by M_3.
Then the points M_1, M_2, M_3 are collinear (cf. Figure 7.6c).

In the following we show that the treatment due to [128, 127] and [23, 124] can be trivially implemented in SINGULAR code:

```
LIB "nctools.lib"; option(redSB); option(redTail);

proc getSubdivision(S, E, s, e)
{
  return ((s*S+e*E)/(s+e));
};

proc getMiddle(S, E)
{
  return (getSubdivision(S,E,1,1));
};

proc getReflection(C, M)
{
  return (2*C - M);
};

proc areEqual(x, y)
{
  return ((x-y));
};

proc isMiddle(x, S, E)
{
  return (areEqual(x, getMiddle(S, E)));
}

proc areCollinear(x, S, E)   // Are x S and x E collinar?
{
  return ((x-S)*(x-E));
};

proc areCoplanar(P1, P2, P3, P4)
{
  return ( (P1 - P2) * (P1 - P2) * (P1 - P3) );
}

proc isParallelogram(P1, P2, P3, P4)
{
  return (P2 - P1 + P4 - P3);
```

```
}
proc areParallel(S1, E1, S2, E2)
{
  return ( (S1 - E1)*(S2 - E2) );
}

proc th(ideal conditions)
"Find all possible implications"
{
  return ( groebner(conditions) ); // no need in twostd, due to homogenity!
}

proc verify(ideal conditions, list #)
"Tests radical ideal membership"
{
  if( size(#) == 1 )
  {
    def statement = #[1];
  }
  else
  {
    def statement = #;
  }

  if( typeof(statement) == "poly" )
  {
    if( statement == 0 )
    {
      return ( list(1==1, 0) );
    }

    def p = statement; int i = 1;

    while( p != 0 )
    {
      if( NF(p, (conditions)) == 0 )
      {
        break;
      }
      i++;
      p = p * statement;
    };

    return ( list( (p != 0), i )  );
  }
  else
  {
    int r = 1; list R, L;

    for( int k = size(statement); k > 0; k-- )
```

7.2. COORDINATE-FREE VERIFICATION OF AFFINE GEOMETRY THEOREMS

```
    {
      L = verify( conditions, statement[k] );
      r = r and L[1];
      R[k] = L[2];
    }

    return (list(r, R));
  }
}
proc areCoplanarByMonom(poly m, int d)
"Auxiliary procedure"
{
  intvec e = leadexp(m);

  int k = size(e);

  while (e[k] == 0)
  {
    k-- ;
  }

  def p = var(k); d--;

  poly r = 1;

  for (k--; (k > 0) and (d > 0); k-- )
  {
    if( e[k] > 0 )
    {
      r = (var(k) - p) * r;
      d--;
    }
  }

  return (r);
}
proc sp(int d)
"Auxiliary procedure"
{
  ideal m = simplify(NF(maxideal(d + 2), std(0)), 2);

  for( int k = size(m); k > 0; k-- )
  {
    m[k] = areCoplanarByMonom(m[k], d + 2);
  }

  return (m);
}
```

```
proc analyzeTheoremK( conditions )
"Get the dimension k"
{
  for( int i = 0; i <= nvars(basering); i++ )
  {
    if( verify( conditions, sp(i) )[1] )
    {
      return (i);
    }
  }

  ERROR("Something went wrong... Bad Theorem?");
}

proc analyzeTheoremD( conditions, k, list #)
"Get the dimension d"
{
  for( int i = k; i >= 0; i-- )
  {
    if( verify( th(conditions, sp(i)), # )[1] )
    {
      return (i);
    }
  }

  ERROR("Something went wrong... Bad Theorem?");
}

proc analyzeTheorem( conditions, list #)
"Full analysis of the theorem given by input"
{
  if( attrib(conditions, "isSB") != 1 )
  {
    conditions = th( conditions );
  }

  list r = verify( conditions, # );

  if( r[1] )
  {
    int k = analyzeTheoremK(conditions);
    int d = analyzeTheoremD(conditions, k, #);

    return ("Theorem is generally true ("+string(r[2])+"), k: " + string(k) + ", d: " +
      string(d) + " (v: " + string(vdim(conditions)) + ")");
  }
  else
  {
    return ("Theorem may not be generally true ("+string(r[2])+", v: " +
      string(vdim(conditions)) + ")");
  }
```

7.2. COORDINATE-FREE VERIFICATION OF AFFINE GEOMETRY THEOREMS

```
}

// Example: Centroid:

ring R = 0,(P,A,B,C),dp; def E = Exterior(); setring E;

def A1 = getMiddle(B, C); def B1 = getMiddle(A, C); def C1 = getMiddle(A, B);

ideal H = areCollinear(P, B, B1), areCollinear(P, C, C1);
def h3   = areCollinear(P, A, A1);

analyzeTheorem(th(H), h3); // H => h3?
↦ Theorem is generally true (1), k: 2, d: 2 (v: 10)

kill E, R; // Example 8: Gauss line

ring R = 0,(A(1..3), B(1..3)), dp; def E = Exterior(); setring E;

// complete quadrilateral:
ideal H =
        // A(3) is the intersection of B1 B2 and A1 A2
        areCollinear(A(1), A(2), A(3)), areCollinear(B(1), B(2), A(3)),
        // B(3) is the intersection of B1 A2 and A1 B2
        areCollinear(A(1), B(2), B(3)), areCollinear(B(1), A(2), B(3));

def M1 = getMiddle(A(1), B(1)); def M2 = getMiddle(A(2), B(2));
def M3 = getMiddle(A(3), B(3));

def h = areCollinear(M1, M2, M3);

analyzeTheorem(th(H), h); // H => h?
↦ Theorem is generally true (1), k: 2, d: 2 (v: 24)

kill E, R; // Example: Pappus' line Theorem

ring R = 0,(A(1..2),B(1..2),C(1..2),M(1..3)),dp;def E = Exterior();setring E;

ideal H = areCollinear(A(1), B(1), C(1)),areCollinear(A(2), B(2), C(2)),
        areCollinear(A(1), M(1), B(2)),areCollinear(A(1), M(2), C(2)),
        areCollinear(A(2), M(1), B(1)),areCollinear(A(2), M(2), C(1)),
        areCollinear(B(1), M(1), A(2)),areCollinear(B(1), M(3), C(2)),
        areCollinear(B(2), M(1), A(1)),areCollinear(B(2), M(3), C(1)),
        areCollinear(C(1), M(2), A(2)),areCollinear(C(1), M(3), B(2)),
        areCollinear(C(2), M(2), A(1)),areCollinear(C(2), M(3), B(1));

def h =   areCollinear(M(1), M(2), M(3));

analyzeTheorem(th(H), h); // H => h?
↦ Theorem may not be generally true (2, v: 74)

kill E, R; // Additional Example: Iterative Reflection
```

```
ring R = 0,(O(1..3), M), dp; def E = Exterior(); setring E;

def M1 = getReflection(O(1), M); def M2 = getReflection(O(2), M1);
def M3 = getReflection(O(3), M2);def M4 = getReflection(O(1), M3);
def M5 = getReflection(O(2), M4);

def h = isMiddle(O(3), M, M5);

analyzeTheorem(th(0), h); // is M5 the reflection of M wrt O(3)?
↦ Theorem is generally true (0), k: 3, d: 3 (v: 16)
```

As our computation shows the Pappus'line Theorem cannot be proved using this approach.

7.3 Super-symmetry

The super-commutative algebras were first introduced as symmetric algebras of super-manifolds in theoretical physics, or more precisely, super-symmetry, which is a part of the theory of elementary particles and their interactions (cf. [92, 13, 93]), where these algebras enable one to join particles with Bose-Einstein statistics and Fermi-Dirac statistics into single multiplets, and also enables one to join the internal and dynamic symmetries of gauge theories in a single super-group.

In particle physics, **super-symmetry** is a symmetry that relates elementary particles of one spin to other particles that differ by half a unit of spin and are known as super-partners. In a theory with unbroken super-symmetry, for every type of boson there exists a corresponding type of fermion with the same mass and internal quantum numbers, and vice-versa.

Remark 7.3.1. As of 2010 there is only indirect evidence (e.g. via *Gauge Coupling Unification*) that super-symmetry may be a symmetry of nature.

The only unambiguous way to claim discovery of super-symmetry is to produce super-particles in the laboratory. Because super-particles are expected to be 100 to 1000 times heavier than the proton, it requires a huge amount of energy to make these particles that can only be achieved at particle accelerators.

Recently physicists have become concerned about the non-discovery of the Higgs boson or any super-partner. Many nevertheless hold out hope on account of the possibility that the Large Hadron Collider, which began operation at CERN in 2009 will discover it.

Super-symmetry is also sometimes studied mathematically for its intrinsic properties. This is because it describes complex fields satisfying a property known as holomorphy, which allows holomorphic quantities to be exactly computed. This makes super-symmetric models useful toy models of more realistic theories. A prime example of this has been the

7.3. SUPER-SYMMETRY

demonstration of S-duality in four dimensional gauge theories that interchanges particles and monopoles.

Super-symmetry has an advantage that super-symmetric quantum field theory can sometimes be solved.

Super-string theory is an attempt to explain all of the particles and fundamental forces of nature in one theory by modeling them as vibrations of tiny super-symmetric strings. Super-string theory is a shorthand for *super-symmetric string theory* because unlike bosonic string theory, it is the version of string theory that incorporates fermions and super-symmetry.

Super-symmetry is also a feature of most versions of string theory, though it can exist in nature even if string theory is incorrect.

Super-algebras and their representations, super-modules, provide an algebraic framework for formulating super-symmetry. The study of such objects is sometimes called super linear algebra. Super-algebras also play an important role in related field of super-geometry, where they enter into the definitions of graded manifolds, super-manifolds and super-schemes.

In theoretical physics, super-gravity theory is a field theory that combines the principles of super-symmetry and general relativity. Together, these imply that, in super-gravity, the super-symmetry is a local symmetry (in contrast to non-gravitational super-symmetric theories, such as the Minimal Super-symmetric Standard Model).

Following [92, 125, 13] we will briefly recall the very basic $\mathbb{Z}_2$-graded mathematical super-notions, originating from physical super-symmetry theory.

- A *super vector space* V is a $\mathbb{Z}_2$-graded[2] vector space $V = V_0 \oplus V_1$ over a field $\Bbbk$, where the grading is defined by putting $|x| = \bar{0}$ for (even) vectors $x \in V_0$ and $|\xi| = \bar{1}$ for (odd) vectors $\xi \in V_1$. If $d_i = \text{Dim}_\Bbbk V_i$ we say that V has *dimension* $d_0 \mid d_1$. For super vector spaces V, W, the morphisms from V to W are $\Bbbk$-linear maps $V \to W$ that preserve the grading. They form a linear space denoted by $\text{Hom}(V, W)$.
- Let us take $V = \Bbbk^{p+q}$ with its standard basis $e_i (1 \leqslant i \leqslant p+q)$ and define e_i to be even (resp. odd) if $i \leqslant p$ (resp., $i > p$), then V becomes a super vector space with

$$V_0 = \sum_{i=1}^{p} \Bbbk \cdot e_i, \quad V_1 = \sum_{i=p+1}^{p+q} \Bbbk \cdot e_i,$$

which we denote by $\Bbbk^{p|q}$.
- Let us denote by $\textbf{Hom}(V, W)$ the vector space of *all* $\Bbbk$-linear maps from V to W, where even maps are the ones preserving grading while odd maps are those that reverse it. In particular, $(\textbf{Hom}(V, W))_0 = \text{Hom}(V, W)$. $\textbf{Hom}(V, W)$ is the so-called *internal* Hom. Denote $\textbf{End}(V) := \textbf{Hom}(V, V)$.

---

[2] where $\mathbb{Z}_2 = \mathbb{Z}/2\mathbb{Z} = \{\bar{0}, \bar{1}\}$

- A ***super-algebra*** $\mathcal{A}$ is a super vector space $\mathcal{A}$ endowed with an (even) associative bilinear product $\mathcal{A} \otimes \mathcal{A} \to \mathcal{A} : a \otimes b \mapsto a * b$, satisfying $|a * b| = |a| +_{\mathbb{Z}_2} |b|$, which is unital.
  In a super-algebra one defines the ***super-commutator*** by $[b, a] := b*a - (-1)^{|a||b|} a*b$, for homogeneous elements a, b.
- If V is a super vector space then $\mathbf{End}(V)$ is a super-algebra.
- A ***super-commutative algebra*** $\mathcal{A}$ is a super-algebra $\mathcal{A}$, where the product satisfies $b * a = (-1)^{|a||b|} a * b$, for homogeneous elements $a, b \in \mathcal{A}$. Equivalently, it is a super-algebra where the super-commutator always vanishes.
- The ***super-center*** of any super-algebra $\mathcal{A}$ is the set of elements that super-commute with all elements, denoted by $\mathbf{Z}(\mathcal{A})$. It is clearly a super-commutative algebra. Clearly $\mathbf{Z}(\mathbf{End}(V)) = \Bbbk \cdot 1$.
- If $V = \Bbbk^{p|q}$ we write $M(p \mid q)$ or $M^{p|q}$ for $\mathbf{End}(V)$. Using the standard basis we have the usual matrix representations for elements of $M(p \mid q)$ in the form
$$\begin{pmatrix} A & B \\ C & D \end{pmatrix},$$
where the letters A, B, C, D denote matrices of orders respectively $p \times p, p \times q, q \times p, q \times q$. The even elements and odd elements are, respectively, of the form
$$\begin{pmatrix} A & 0 \\ 0 & D \end{pmatrix}, \begin{pmatrix} 0 & B \\ C & 0 \end{pmatrix}.$$

- Let V be a finite-dimensional super vector space and let $X \in \mathbf{End}(V)$. Then we have
$$X = \begin{pmatrix} X_{00} & X_{01} \\ X_{10} & X_{11} \end{pmatrix},$$
where X_{ij} is the linear map from V_j to V_i such that $X_{ij}v$ is the projection on V_i of Xv for $v \in V_j$.
The ***super-trace*** of X is defined as
$$\mathrm{sTr}(X) := \mathrm{Tr}(X_{00}) - \mathrm{Tr}(X_{11}) \in \Bbbk.$$

The ***Berezinian*** of X is defined as
$$\mathrm{Ber}(X) := \mathrm{Det}(X_{00}) \mathrm{Det}\left(1 - X_{01} X_{11}^{-1} X_{10}\right) \mathrm{Det}(X_{11})^{-1} \in \Bbbk.$$

It has the important property that
$$\mathrm{Ber}(XY) = \mathrm{Ber}(X) \mathrm{Ber}(Y), \ X, Y \in \mathbf{End}(V).$$

7.3. SUPER-SYMMETRY

- Let $\mathcal{A}$ be a super-commutative $\Bbbk$-algebra. Axioms of left-, right- and bi- $\mathcal{A}$-modules coincide with $\mathbb{Z}_2$-graded versions of the usual axioms, e.g. there are no additional changes in sign. Moreover, any one-sided module can be canonically considered as a bi-module. Since the same holds for ideals in $\mathcal{A}$ we will only consider left ideals.
  Left modules are super $\Bbbk$-vector spaces on which $\mathcal{A}$ acts from the left; and the action is a morphism of super vector spaces:

$$a \otimes m \mapsto a \cdot m, |a \cdot m| = |a| + |m|, a \in \mathcal{A}, m \in M.$$

As in the classical theory, left modules may be viewed as right modules and *vice versa*, but in the super case this involves sing factors; thus a left module M is viewed as a right module for $\mathcal{A}$ under the action

$$m \cdot a = (-1)^{|a||m|} a \cdot m, a \in \mathcal{A}, m \in M.$$

A morphism $\Psi : M \to N$ of (super) $\mathcal{A}$-modules is an even $\Bbbk$-linear map such that $\Psi(am) = a\Psi(m)$.
For modules M, N one has $M \otimes N$ defined in the usual manner by dividing $M \otimes_{\Bbbk} N$ by the $\Bbbk$-linear spanned by the relations $ma \otimes n = m \otimes an, a \in \mathcal{A}$.
We denote by $\mathbf{Hom}(M, N)$ the space if *all* $\Bbbk$-linear maps $\Psi : M \to N$ such that $\Psi(am) = (-1)^{|\Psi||a|} a\Psi(m)$. It is called the internal Hom.
Clearly $\mathbf{Hom}(M, N)$ is again an $\mathcal{A}$-module if we define $(a\Psi)(m) := a\Psi(m)$.
Denote the module dual to M by $M^\vee := \mathbf{Hom}(M, \mathcal{A})$ and $\mathbf{End}(M) := \mathbf{Hom}(M, M)$.

- A *free (super)* $\mathcal{A}$*-module* is an $\mathcal{A}$-module that has a free *homogeneous* basis. If $e_i (1 \leqslant i \leqslant p+q)$ is a basis with e_i even or odd according as $i \leqslant p$ or $p+1 \leqslant i \leqslant p+q$, we denote it by $\mathcal{A}^{p|q}$, and define its rank as $p \mid q$. Thus

$$\mathcal{A}^{p|q} = \underbrace{(\mathcal{A}e_1 \oplus \cdots \oplus \mathcal{A}e_p)}_{\text{even}} \oplus \underbrace{(\mathcal{A}e_{p+1} \oplus \cdots \oplus \mathcal{A}e_{p+q})}_{\text{odd}}.$$

- Let $\mathcal{T}(V)$ be the tensor algebra of a super vector space V. Let us denote by I a two-sided ideal in $\mathcal{T}(V)$ which is generated by the following elements: $b \otimes a - (-1)^{|a||b|} a \otimes b$ for all $a, b \in V$. We define the symmetric algebra of V to be the factor algebra $\mathrm{Sym}(V) := \mathcal{T}(V)/I$. Obviously $\mathcal{T}(V)$ inherits a $\mathbb{Z}_2$-grading and so does the factor by the $\mathbb{Z}_2$-graded ideal I.
  In a sense the symmetric algebra $\mathrm{Sym}(V)$ is a super-commutative algebra of algebraic (polynomial) functions on an affine super vector space V.
  Moreover, one has: $\mathrm{Sym}(V) = \mathrm{Sym}(V)_0 \oplus \mathrm{Sym}(V)_1 = \mathcal{T}(V)_0/I_0 \oplus \mathcal{T}(V)_1/I_1$. Hence if $V = V_0$ then $\mathrm{Sym}(V)$ is the usual symmetric algebra on V_0 (which is a commutative polynomial algebra) and if $V = V_1$ then $\mathrm{Sym}(V)$ is the exterior algebra, which is clearly $\mathbb{Z}_2$-graded and thus super-commutative.

- The concept of a manifold generalizes so that the functions can take values in a commutative super-algebra.
  The structure of a super-manifold on a manifold M with a structure sheaf $\mathcal{O}_M$ is defined by a sheaf of commutative super-algebras $\mathcal{F}$ over the sheaf $\mathcal{O}_M$, whereby

any point $p \in M$ possesses a neighborhood U such that the ringed space $(U, \mathcal{F}|_U)$ is isomorphic to $(U, (\mathcal{O}_M|_U) \otimes \bigwedge(\mathbb{A}^m))$, where $\bigwedge(\mathbb{A}^m)$ is the exterior algebra with m odd generators. The pair $\operatorname{Dim} M \mid m$ is called the dimension of the super-manifold. Analytic super-manifolds are defined in the same way. The differentiable (or analytic) super-manifolds form a category whose morphisms are the morphisms of ringed spaces that are even on the structure sheaves.

A super-manifold of the form $(U, \mathcal{O}_U \otimes \bigwedge(\mathbb{A}^m))$, where $(U, \mathcal{O}_U)$ is an open submanifold in $\mathbb{A}^n$, is called a super-domain of dimension $n \mid m$. Note that every super-manifold is locally isomorphic to a super-domain.

7.4 Cohomology rings of finite p-groups

Let p be a prime integer and G be a finite p-group (i.e., $|G| = p^r$). The cohomology ring $\mathrm{H}^*(G; \mathbf{F}_p)$ is graded commutative finitely presented $\mathbf{F}_p$-algebra, determined by G up to an isomorphism.

Cohomology rings of finite p-groups with coefficients in $\mathbf{F}_p$ can be computed by constructing degree-wise approximations of the cohomology ring until approximation is isomorphic to the actual cohomology ring (which can be checked using the Benson's completeness criterion). This basic approach (which is due to Jon Carlson) The recent implementation of the approach by David Green and Simon King (in the DFG project **GR 1585/4-1**: http://users.minet.uni-jena.de/~king/cohomology/index.html) is based on the broad range of free open source Computer Algebra software that is accessible with the free open-source mathematics software system SAGE (cf. [121]), which, in particular, includes GAP (cf. [54]) and SINGULAR.

They have successfully used SINGULAR for all computations in graded commutative algebras (i.e. the SINGULAR kernel framework, which has been developed within and due to this thesis): computing a Gröbner basis of the relation ideal, detecting relations, partially for constructing simultaneous lifts of the subgroup Dickson invariants, for detecting filter-regular systems of parameters and for computing their filter degree type.

For further information we refer the interested reader to [62], http://users.minet.uni-jena.de/~king/cohomology/background.html, testing Benson's regularity conjecture: http://www.math.rwth-aachen.de:8001/Nikolaus2007/abstracts/green2007.pdf, the Cohomology of finite p-Groups: http://hamilton.nuigalway.ie/DeBrunCentre/SecondWorkshop/simon.pdf.

Chapter 8
Conclusion and Future Work

Using our non-commutative framework it is now easy to develop further extensions, say for Clifford algebras or remove the limitation about a single block of non-commutative variables for SCA.

At the moment the only systems, known to have facilities for computing sheaf cohomologies via BGG are M2 (the very first system to support these computations), SINGULAR (at the beginning via SINGULAR:PLURAL and later via our implementation of graded commutative algebras) and recently Magma and the GAP package HomAlg. The later relies on a CAS such as SINGULAR, M2 or Magma to do the actual computations. Up until now Macaulay2 was the leader in this special computations. Our tests/experiments with sheaf cohomology computations revealed a number of bottlenecks in the SINGULAR implementation by W. Decker, Ch. Lossen and G. Pfister and with our tweaks (in both commutative and non-commutative parts) SINGULAR has become generally faster than Macaulay2 on our set of tests.

Magma has got exterior algebras specially implemented very recently. Unfortunately, our experience with Magma showed that their implementation is still not usable due to segmentation faults. And thus, can not provide timing for Magma at the moment.

At the beginning of our experience with HomAlg (with SINGULAR as a backend), it was quit a way behind both SINGULAR and M2. But due to our cooperation with its authors, esp. with Dr. Mohamed Barakat, it was possible to boost its performance so that it has become the leader. It should be noted that HomAlg has somewhat more general implementation of Tate resolution than we used to benchmark SINGULAR and M2. Moreover, HomAlg computes graded resolutions via iterated syzygies over graded commutative algebras.

Our short experience with HomAlg showed that while being as abstract as possible it can be easily tweaked and obtain excellent performance. In order to implement the algorithm (cf. [39]) for higher direct images of sheaves we need a much better concept of (multi-)gradings in SINGULAR. Therefore it would be better to improve the HomAlg's SINGULAR-related functionality rather than implement this algorithm in SINGULAR script language, which is rather inconvenient for such abstract designs.

Research and implement more involved algorithms for syzygies and resolutions over graded commutative algebras using the developed framework (our generalized Schreyer ordering in SINGULAR).

We have developed a framework for experiments with Schreyer induced orderings in general non-commutative setting and prototype implementation of Algorithm 5.4.3 which outperformed the SINGULAR's kernel implementation on several tests, despite being written in Singular script language. Further we are going to experiment with the advanced free resolution Algorithm 5.4.4 using our framework.

We are further interested in investigating relations between our approach to syzygy and Schreyer resolution computations and the results (specific criteria) from [47] and [20].

Bibliography

[1] ADAMS, W. W., AND LOUSTAUNAU, P. *An introduction to Gröbner bases.* Graduate Studies in Mathematics. 3. Providence, RI: American Mathematical Society (AMS)., 1994.

[2] ANDERSON, F. W., AND FULLER, K. R. *Rings and categories of modules,* second ed., vol. 13 of *Graduate Texts in Mathematics.* Springer-Verlag, New York, 1992.

[3] ANICK, D. J. Non-commutative graded algebras and their Hilbert series. *J. Algebra 78* (1980), 120–140.

[4] ANICK, D. J. On the homology of associative algebras. *Trans. Amer. Math. Soc. 296,* 2 (1986), 641–659.

[5] APEL, J. *Grobnerbasen in nichtkommutativen Algebren und ihre Anwendung.* Dissertation, University of Leipzig, Leipzig, 1988.

[6] APEL, J. Computational ideal theory in finitely generated extension rings. *Theor. Comput. Sci. 244,* 1-2 (2000), 1–33.

[7] BACHMANN, O., AND GRÄBE, H.-G. The SYMBOLICDATA Project: Towards an Electronic Repository of Tools and Data for Benchmarks of Computer Algebra Software. In *Reports On Computer Algebra,* no. 27. Centre for Computer Algebra, University of Kaiserslautern, Jan 2000. http://www.mathematik.uni-kl.de/~zca.

[8] BACHMANN, O., AND SCHÖNEMANN, H. Monomial representations of Gröbner bases computations. Gloor, Oliver (ed.), Proceedings of the 1998 international symposium on symbolic and algebraic computation, ISSAC '98, Rostock, Germany, August 13–15, 1998. New York, NY: ACM Press. 309-316 (1998)., 1998.

[9] BARAKAT, M., AND ROBERTZ, D. The homalg project – algorithmic homological algebra. http://homalg.math.rwth-aachen.de/.

[10] BECKER, T., AND WEISPFENNING, V. *Gröbner bases: a computational approach to commutative algebra. In cooperation with Heinz Kredel.* Graduate Texts in Mathematics. 141. New York: Springer-Verlag. xxii, 574 p. , 1993.

[11] BEILINSON, A. Coherent sheaves on P^n and problems of linear algebra. *Funct. Anal. Appl. 12* (1979), 214–216.

[12] BEILINSON, A., GINZBURG, V., AND SOERGEL, W. Koszul duality patterns in representation theory. *J. Amer. Math. Soc. 9*, 2 (1996), 473–527.

[13] BEREZIN, F. A. The mathematical basis of supersymmetric field theories. *Soviet J. Nucl. Phys. 29* (1979), 857–866.

[14] BERGMAN, G. M. The diamond lemma for ring theory. *Adv. Math. 29* (1977), 178–218.

[15] BERNSHTEJN, I., GEL'FAND, I., AND GEL'FAND, S. Algebraic bundles over P^n and problems of linear algebra. *Funct. Anal. Appl. 12* (1979), 212–214.

[16] BUCHBERGER, B. *Ein Algorithmus zum Auffinden der Basiselemente des Restklassenringes nach einem nulldimensionalen Polynomideal.* Dissertation, University of Innsbruck, Innsbruck, 1965.

[17] BUCHBERGER, B. A criterion for detecting unnecessary reductions in the construction of Gröbner-bases. In *Symbolic and algebraic computation (EUROSAM '79, Internat. Sympos., Marseille, 1979)*, vol. 72 of *Lecture Notes in Comput. Sci.* Springer, Berlin, 1979, pp. 3–21.

[18] BUCHBERGER, B. *Gröbner-Bases: An Algorithmic Method in Polynomial Ideal Theory.* Reidel Publishing Company, Dodrecht - Boston - Lancaster, 1985.

[19] BUESO, J., GÓMEZ-TORRECILLAS, J., AND VERSCHOREN, A. *Algorithmic methods in non-commutative algebra*, vol. 17 of *Mathematical Modelling: Theory and Applications*. Kluwer Academic Publishers, Dordrecht, 2003. Applications to quantum groups.

[20] CABOARA, M., KREUZER, M., AND ROBBIANO, L. Efficiently computing minimal sets of critical pairs. *J. Symb. Comput. 38*, 4 (2005), 1169–1190.

[21] CAPANI, A., DOMINICIS, G. D., NIESI, G., AND ROBBIANO, L. Computing minimal finite free resolutions. *J. Pure Appl. Algebra 117/118* (1997), 105–117. Algorithms for algebra (Eindhoven, 1996).

[22] CHOU, S.-C., AND GAO, X.-S. Automated reasoning in geometry. In Robinson and Voronkov [109], pp. 707–749.

[23] CHUPAEVA, I. Automated proving and analysis of geometric theorems in coordinate-free form by using the anticommutative Gröbner basis method. *Fundam. Prikl. Mat. 9*, 3 (2003), 213–228.

BIBLIOGRAPHY

[24] COHN, P. M. *Free rings and their relations*. Academic Press, London, 1971. London Mathematical Society Monographs, No. 2.

[25] COHN, P. M. Rings of fractions. *Amer. Math. Monthly 78* (1971), 596–615.

[26] COJOCARU, S., PODOPLELOV, A., AND UFNAROVSKI, V. Non-commutative gröbner bases and anick's resolution. Dräxler, P. (ed.) et al., Computational methods for representations of groups and algebras. Proceedings of the Euroconference in Essen, Germany, April 1-5, 1997. Basel: Birkhäuser. Prog. Math. 173, 139-159 (1999)., 1999.

[27] COX, D., LITTLE, J., AND O'SHEA, D. *Ideals, varieties, and algorithms. An introduction to computational algebraic geometry and commutative algebra*. Undergraduate Texts in Mathematics. New York: Springer-Verlag. xi, 513 p. , 1992.

[28] DECKER, W., EIN, L., AND SCHREYER, F.-O. Construction of surfaces in P^4. *J. Algebr. Geom. 2*, 2 (1993), 185–237.

[29] DECKER, W., AND EISENBUD, D. Sheaf algorithms using the exterior algebra. In *Computations in algebraic geometry with Macaulay 2* [44], pp. 215–249.

[30] DECKER, W., AND LOSSEN, C. *Computing in algebraic geometry. A quick start using SINGULAR*. Algorithms and Computation in Mathematics 16. Berlin: Springer; New Delhi: Hindustan Book Agency. xvi, 327 p., 2006.

[31] DECKER, W., LOSSEN, C., PFISTER, G., AND MOTSAK, O. sheafcoh.lib. a SINGULAR library for computing sheaf cohomology.

[32] DECKER, W., AND SCHREYER, F.-O. Varieties, groebner bases, and algebraic curves. To appear.

[33] DICKSON, L. E. Finiteness of the Odd Perfect and Primitive Abundant Numbers with n Distinct Prime Factors. *Amer. J. Math. 35*, 4 (1913), 413–422.

[34] EISENBUD, D. *Commutative algebra. With a view toward algebraic geometry*. Graduate Texts in Mathematics. 150. Berlin: Springer-Verlag. xvi, 785 p. , 1995.

[35] EISENBUD, D. An exterior view of modules and sheaves. In *Advances in algebra and geometry (Hyderabad, 2001)*. Hindustan Book Agency, New Delhi, 2003, pp. 209–216.

[36] EISENBUD, D., FLØYSTAD, G., AND SCHREYER, F.-O. Sheaf cohomology and free resolutions over exterior algebras. *Trans. Am. Math. Soc. 355*, 11 (2003), 4397–4426.

[37] EISENBUD, D., POPESCU, S., SCHREYER, F.-O., AND WALTER, C. Exterior algebra methods for the minimal resolution conjecture. *Duke Math. J. 112*, 2 (2002), 379–395.

[38] EISENBUD, D., POPESCU, S., AND YUZVINSKY, S. Hyperplane arrangement cohomology and monomials in the exterior algebra. *Trans. Am. Math. Soc. 355*, 11 (2003), 4365–4383.

[39] EISENBUD, D., AND SCHREYER, F.-O. Relative Beilinson Monad and direct image for families of coherent sheaves. *Trans. Am. Math. Soc. 360*, 10 (2008), 5367–5396.

[40] EISENBUD, D., AND SCHREYER, F.-O. Betti numbers of graded modules and cohomology of vector bundles. *J. Amer. Math. Soc. 22*, 3 (2009), 859–888.

[41] EISENBUD, D., AND SCHREYER, F.-O. Cohomology of coherent sheaves and series of supernatural bundles. http://www.citebase.org/abstract?id=oai:arXiv.org:0902.1594, 2009.

[42] EISENBUD, D., SCHREYER, F.-O., AND WEYMAN, J. Resultants and Chow forms via exterior syzygies. *J. Amer. Math. Soc. 16*, 3 (2003), 537–579 (electronic).

[43] EISENBUD, D., AND WEYMAN, J. Fitting's lemma for $\mathbb{Z}/2$-graded modules. *Trans. Am. Math. Soc. 355*, 11 (2003), 4451–4473.

[44] EISENBUD, D. E., GRAYSON, D. R. E., STILLMAN, M. E., AND STURMFELS, B. E. *Computations in algebraic geometry with Macaulay 2*. Algorithms and Computation in Mathematics. 8. Berlin: Springer. xiii, 329 p. , 2002.

[45] EL FROM, Y. On the algebras of solvable type. (Sur les algèbres de type résoluble.). *Afrika Mat. (3) 4* (1994), 1–10.

[46] FAUGÈRE, J.-C. A new efficient algorithm for computing Gröbner bases (F_4). *J. Pure Appl. Algebra 139*, 1-3 (1999), 61–88.

[47] FAUGÈRE, J.-C. A new efficient algorithm for computing Gröbner bases without reduction to zero (F_5). Mora, Teo (ed.), ISSAC 2002. Proceedings of the 2002 international symposium on symbolic and algebraic computation, Lille, France, July 07–10, 2002. New York, NY: ACM Press. 75-83 (2002)., 2002.

[48] FEARNLEY-SANDER, D. Affine geometry and exterior algebra. *Houston J. Math. 6* (1980), 53–58.

[49] FEARNLEY-SANDER, D. Hermann Grassmann and the prehistory of universal algebra. *Am. Math. Mon. 89* (1982), 161–166.

[50] FEARNLEY-SANDER, D. Plane Euclidean reasoning. Gao, Xiao-Shan (ed.) et al., Automated deduction in geometry. 2nd international workshop, ADG '98, Beijing, China, August 1-3, 1998. Proceedings. Berlin: Springer. Lect. Notes Comput. Sci. 1669, 86-110 (1999)., 1999.

[51] FEARNLEY-SANDER, D., AND STOKES, T. Area in Grassmann geometry. In *Automated deduction in geometry. International workshop, Toulouse, France, September 27–29, 1996. Proceedings.* [127].

[52] FLOYSTAD, G. Describing coherent sheaves on projective spaces via koszul duality. http://www.citebase.org/abstract?id=oai:arXiv.org:math/0012263, 2000.

[53] FLØYSTAD, G. Koszul duality and equivalences of categories. *Trans. Amer. Math. Soc. 358*, 6 (2006), 2373–2398 (electronic).

[54] THE GAP GROUP. *GAP – Groups, Algorithms, and Programming*, 2008.

[55] GARCÍA ROMÁN, M., AND GARCÍA ROMÁN, S. Gröbner bases and syzygies on bimodules over PBW algebras. *J. Symb. Comput. 40*, 3 (2005), 1039–1052.

[56] GEBAUER, R., AND MÖLLER, H. On an installation of Buchberger's algorithm. *J. Symb. Comput. 6*, 2-3 (1988), 275–286.

[57] GELFAND, S., AND MANIN, Y. *Methods of homological algebra. Transl. from the Russian. 2nd ed.* Springer Monographs in Mathematics. Berlin: Springer. xx, 372 p., 2003.

[58] GIANNI, P. M., Ed. *Symbolic and Algebraic Computation, International Symposium ISSAC'88, Rome, Italy, July 4-8, 1988, Proceedings* (1989), vol. 358 of *Lecture Notes in Computer Science*, Springer.

[59] GOODEARL, K. R., AND WARFIELD, R. *An introduction to noncommutative Noetherian rings.* London Mathematical Society Student Texts, 16. Cambridge ect.: Cambridge University Press. xvii, 303 p., 1989.

[60] GRASSMANN, H. *Die Ausdehnungslehre.* Berlin, 1862.

[61] GRAYSON, D. R., AND STILLMAN, M. E. Macaulay2, a software system for research in algebraic geometry. Available at http://www.math.uiuc.edu/Macaulay2/.

[62] GREEN, D. J. *Gröbner bases and the computation of group cohomology.* Lecture Notes in Mathematics 1828. Berlin: Springer. xi, 138 p., 2003.

[63] GREEN, E., MORA, T., AND UFNAROVSKI, V. The non-commutative Gröbner freaks. In *Symbolic rewriting techniques (Ascona, 1995)*, vol. 15 of *Progr. Comput. Sci. Appl. Logic.* Birkhäuser, Basel, 1998, pp. 93–104.

[64] GREEN, E. L. Noncommutative gröbner bases, and projective resolutions. Dräxler, P. (ed.) et al., Computational methods for representations of groups and algebras. Proceedings of the Euroconference in Essen, Germany, April 1-5, 1997. Basel: Birkhäuser. Prog. Math. 173, 29-60 (1999)., 1999.

[65] GREEN, E. L. Multiplicative bases, gröbner bases, and right gröbner bases. *J. Symb. Comput. 29*, 4-5 (2000), 601–623.

[66] GREUEL, G.-M., AND PFISTER, G. Advances and improvements in the theory of standard bases and syzygies. *Arch. Math. 66*, 2 (1996), 163–176.

[67] GREUEL, G.-M., AND PFISTER, G. *A Singular introduction to commutative algebra*. With contributions by Olaf Bachmann, Christoph Lossen and Hans Schönemann. 2nd extended ed. Berlin: Springer. xx, 689 p. , 2007.

[68] GREUEL, G.-M., PFISTER, G., AND SCHÖNEMANN, H. SINGULAR 3.0. A Computer Algebra System for Polynomial Computations, Centre for Computer Algebra, University of Kaiserslautern, 2005. http://www.singular.uni-kl.de.

[69] GROSSHANS, F. D., ROTA, G.-C., AND STEIN, J. A. *Invariant theory and superalgebras*, vol. 69 of *CBMS Regional Conference Series in Mathematics*. Published for the Conference Board of the Mathematical Sciences, Washington, DC, 1987.

[70] HARTLEY, D., AND TUCKER, R. W. A constructive implementation of the Cartan-Kähler theory of exterior differential systems. *J. Symbolic Comput. 12*, 6 (1991), 655–667.

[71] HARTLEY, D., AND TUCKEY, P. Xideal, gröbner bases for exterior algebra (reduce library package), 1993. http://www.reduce-algebra.com/docs/xideal.pdf.

[72] HARTLEY, D., AND TUCKEY, P. Gröbner bases in clifford and grassmann algebras. *J. Symb. Comput. 20*, 2 (1995), 197–205.

[73] HARTSHORNE, R. *Algebraic geometry*. Graduate Texts in Mathematics. 52. New York-Heidelberg-Berlin: Springer-Verlag. XVI, 496 p. , 1977.

[74] HEYWORTH, A. One-sided noncommutative gröbner bases with applications to computing green's relations. *J. Algebra 242*, 2 (2001), 401–416.

[75] JØRGENSEN, P. Linear free resolutions over non-commutative algebras. *Compos. Math. 140*, 4 (2004), 1053–1058.

[76] KANDRI-RODY, A., AND WEISPFENNING, V. Noncommutative Gröbner bases in algebras of solvable type. *J. Symbolic Comput. 9*, 1 (1990), 1–26.

[77] KOBAYASHI, Y. Gröbner bases of associative algebras and the hochschild cohomology. *Trans. Am. Math. Soc. 357*, 3 (2005), 1095–1124.

[78] KREDEL, H. *Solvable Polynomial Rings*. Dissertation, Universität Passau, Passau, 1992.

[79] LA SCALA, R. *A computational approach to minimal free resolutions*. Ph.d, University of Pisa, 1994. Draft version in English, via private communication.

BIBLIOGRAPHY 145

[80] LA SCALA, R., AND LEVANDOVSKYY, V. Letterplace ideals and non-commutative Gröbner bases. *J. Symbolic Comput. 44*, 10 (2009), 1374–1393.

[81] LA SCALA, R., AND STILLMAN, M. Strategies for computing minimal free resolutions. *J. Symb. Comput. 26*, 4 (1998), 409–431.

[82] LAM, T. *Lectures on modules and rings.* Graduate Texts in Mathematics. 189. New York, NY: Springer. xxiii, 557 p. , 1999.

[83] LAM, T. *A first course in noncommutative rings. 2nd ed.* Graduate Texts in Mathematics. 131. New York, NY: Springer. xix, 385 p. , 2001.

[84] LANG, S. *Algebra. 3rd revised ed.* Graduate Texts in Mathematics. 211. New York, NY: Springer. xv, 914 p. , 2002.

[85] LAZARD, D. Gröbner bases, Gaussian elimination and resolution of systems of algebraic equations. Computer algebra, EUROCAL '83, Proc. Conf., London 1983, Lect. Notes Comput. Sci. 162, 146-156 (1983)., 1983.

[86] LEVANDOVSKYY, V. *Non-commutative Computer Algebra for polynomial algebras: Gröbner bases, applications and implementation.* Dissertation, Technische Universität Kaiserslautern, Kaiserslautern, 2005.

[87] LEVANDOVSKYY, V. PBW bases, non-degeneracy conditions and applications. Buchweitz, Ragnar-Olaf (ed.) et al., Representations of algebras and related topics. Proceedings from the 10th international conference, ICRA X, Toronto, Canada, July 15–August 10, 2002. Dedicated to V. Dlab on the occasion of his 70th birthday. Providence, RI: American Mathematical Society (AMS). Fields Institute Communications 45, 229-246 (2005)., 2005.

[88] LEVANDOVSKYY, V., LOBILLO, F., RABELO, C., AND MOTSAK, O. nctools.lib. SINGULAR library: general tools for noncommutative algebras.

[89] LI, H. *Noncommutative Gröbner bases and filtered-graded transfer.* Lecture Notes in Mathematics 1795. Berlin: Springer. ix, 197 p. , 2002.

[90] MAC LANE, S. *Homology.* Die Grundlehren der mathematischen Wissenschaften, Bd. 114. Academic Press Inc., Publishers, New York, 1963.

[91] MADLENER, K., AND REINERT, B. Computing Gröbner bases in monoid and group rings. Bronstein, Manuel (ed.), ISSAC '93. Proceedings of the 1993 international symposium on Symbolic and algebraic computation, Kiev, Ukraine, July 6–8, 1993. Baltimore, MD: ACM Press. 254-263 (1993)., 1993.

[92] MANIN, Y. I. *Gauge field theory and complex geometry. Transl. from the Russian by N. Koblitz and J. R. King. With an appendix by S. Merkulov. 2nd ed.* Grundlehren der Mathematischen Wissenschaften. 289. Berlin: Springer. xii, 346 p. , 1997.

[93] MARTIN, S. P. A supersymmetry primer. http://www.citebase.org/abstract?id=oai:arXiv.org:hep-ph/9709356, 1997.

[94] MCCONNELL, J., AND ROBSON, J. *Noncommutative Noetherian rings. With the cooperation of L. W. Small. Reprinted with corrections from the 1987 original.* Graduate Studies in Mathematics. 30. Providence, RI: American Mathematical Society (AMS). xx, 636 p., 2001.

[95] MÖLLER, H. M. A reduction strategy for the taylor resolution. In *EUROCAL '85: Research Contributions from the European Conference on Computer Algebra-Volume 2* (1985), Springer-Verlag, pp. 526–534.

[96] MÖLLER, H. M., MORA, T., AND TRAVERSO, C. Gröbner bases computation using syzygies. In *ISSAC* (1992), pp. 320–328.

[97] MORA, F. Gröbner bases for non-commutative polynomial rings. Algebraic algorithms and error-correcting codes, Proc. 3rd Int. Conf., Grenoble/France 1985, Lect. Notes Comput. Sci. 229, 353-362 (1986)., 1986.

[98] MORA, T. Groebner bases in non-commutative algebras. In Gianni [58], pp. 150–161.

[99] MORA, T. An introduction to commutative and noncommutative Gröbner bases. *Theor. Comput. Sci. 134*, 1 (1994), 131–173.

[100] MORIER-GENOUD, S., AND OVSIENKO, V. Simple graded commutative algebras. http://www.citebase.org/abstract?id=oai:arXiv.org:0904.2825, 2009.

[101] NOETHER, E., AND SCHMEIDLER, W. Moduln in nichtkommutativen Bereichen, insbesondere aus Differential- und Differenzenausdrücken. *Math. Z. 8*, 1-2 (1920), 1–35.

[102] NÜSSLER, T., AND SCHÖNEMANN, H. Groebner bases in algebras with zero-divisors. Tech. rep., Fachbereich Mathematik, Universitaet Kaiserslautern, 1993. Preprint 244.

[103] ORE, O. Linear equations in non-commutative fields. *Ann. of Math. (2) 32*, 3 (1931), 463–477.

[104] ORE, O. Theory of non-commutative polynomials. *Ann. of Math. (2) 34*, 3 (1933), 480–508.

[105] PIERCE, R. S. *Associative algebras*, vol. 88 of *Graduate Texts in Mathematics*. Springer-Verlag, New York, 1982. Studies in the History of Modern Science, 9.

[106] PRIDDY, S. B. Koszul resolutions. *Trans. Am. Math. Soc. 152* (1970), 39–60.

[107] RICHMAN, F. Constructive aspects of noetherian rings. *Proc. Am. Math. Soc. 44* (1974), 436–441.

[108] ROBBIANO, L. Term orderings on the polynomial ring. Computer algebra, EUROCAL '85, Proc. Eur. Conf., Linz/Austria 1985, Vol. 2, Lect. Notes Comput. Sci. 204, 513-517 (1985)., 1985.

[109] ROBINSON, J. A., AND VORONKOV, A., Eds. *Handbook of Automated Reasoning (in 2 volumes)*. Elsevier and MIT Press, 2001.

[110] ROTA, G.-C., AND STURMFELS, B. Introduction to invariant theory in superalgebras. Invariant theory and tableaux, Proc. Workshop, Minneapolis/MN (USA) 1988, IMA Vol. Math. Appl. 19, 1-35 (1990)., 1990.

[111] SCHAFER, R. *An introduction to nonassociative algebras*. Pure and Applied Mathematics, 22. A Series of Monographs and Textbooks. New York and London: Academic Press. X, 166 p. , 1966.

[112] SCHÖNEMANN, H. Algorithms in singular. In *Reports On Computer Algebra*, no. 02. Centre for Computer Algebra, University of Kaiserslautern, June 1996.

[113] SCHREYER, F.-O. Die berechnung von syzygien mit dem veralgemeinerten weierstrass'chen divisionssatz. Master's thesis, Univ. Hamburg, 1980.

[114] SCHREYER, F.-O. A standard basis approach to syzygies of canonical curves. *J. Reine Angew. Math. 421* (1991), 83–123.

[115] SERRE, J.-P. Faisceaux algébriques cohérents. *Ann. of Math. (2) 61* (1955), 197–278.

[116] SIEBERT, T. On strategies and implementations for computations of free resolutions. In *Reports On Computer Algebra*, no. 08. Centre for Computer Algebra, University of Kaiserslautern, September 1996.

[117] SIEBERT, T. Recursive Computation of Free Resolutions and a Generalized Koszul Complex. In *Reports On Computer Algebra*, no. 28. Centre for Computer Algebra, University of Kaiserslautern, Jan 2000.

[118] SINGULAR TEAM. SINGULAR online manual. Available at http://www.singular.uni-kl.de/Manual/latest/index.htm.

[119] SMITH, G. G. Computing global extension modules. *J. Symb. Comput. 29*, 4-5 (2000), 729–746.

[120] SPEAR, D. A constructive approach to commutative ring theory. In *Proceedings of the 1977 MACSYMA Users' Conference, NASA CP-2012* (1977), pp. 369–376.

[121] STEIN, W., ET AL. *Sage Mathematics Software*. The Sage Development Team, 2009.

[122] STILLMAN, M. Computing with sheaves and sheaf cohomology in algebraic geometry: preliminary version. http://math.arizona.edu/~swc/aws/notes/files/06StillmanNotes.pdf, February 2006.

[123] STOKES, T. Gröbner bases in exterior algebra. *J. Automat. Reason. 6*, 3 (1990), 233–250.

[124] TCHOUPAEVA, I. Analysis of geometrical theorems in coordinate-free form by using anticommutative Gröbner bases method. Winkler, Franz (ed.), Automated deduction in geometry. 4th international workshop, ADG 2002, Hagenberg Castle, Austria, September 4–6, 2002. Revised papers. Berlin: Springer. Lecture Notes in Computer Science 2930. Lecture Notes in Artificial Intelligence, 178-193 (2004)., 2004.

[125] VARADARAJAN, V. S. *Supersymmetry for mathematicians: an introduction*. Courant Lecture Notes in Mathematics 11. Providence, RI: American Mathematical Society (AMS); New York, NY: Courant Institute of Mathematical Sciences. vi, 300 p. , 2004.

[126] VASCONCELOS, W. V. *Computational methods of commutative algebra and algebraic geometry. With chapters by David Eisenbud, Daniel R. Grayson, Jürgen Herzog and Michael Stillman. 3rd printing*. Algorithms and Computation in Mathematics 2. Berlin: Springer. xiii, 408 p , 2004.

[127] WANG, D. *Automated deduction in geometry. International workshop, Toulouse, France, September 27–29, 1996. Proceedings*. Lecture Notes in Computer Science. Lecture Notes in Artificial Intelligence. 1360. Berlin: Springer. vii, 235 p. , 1998.

[128] WANG, D. Clifford algebraic calculus for geometric reasoning with application to computer vision. In *Automated deduction in geometry. International workshop, Toulouse, France, September 27–29, 1996. Proceedings*. [127].

Die VDM Verlagsservicegesellschaft sucht für wissenschaftliche Verlage abgeschlossene und herausragende

Dissertationen, Habilitationen, Diplomarbeiten, Master Theses, Magisterarbeiten usw.

für die kostenlose Publikation als Fachbuch.

Sie verfügen über eine Arbeit, die hohen inhaltlichen und formalen Ansprüchen genügt, und haben Interesse an einer honorarvergüteten Publikation?

Dann senden Sie bitte erste Informationen über sich und Ihre Arbeit per Email an *info@vdm-vsg.de*.

Sie erhalten kurzfristig unser Feedback!

VDM Verlagsservicegesellschaft mbH
Dudweiler Landstr. 99   Telefon +49 681 3720 174
D - 66123 Saarbrücken   Fax   +49 681 3720 1749
www.vdm-vsg.de

Die VDM Verlagsservicegesellschaft mbH vertritt

Printed by Books on Demand GmbH, Norderstedt / Germany